Strategy and Defense Policy for Small States

Problems and Prospects

Strategy and
Defense Policy
for Small States

Problems and Prospects

edited by

Bernard F W Loo

S. Rajaratnam School of International Studies,
Nanyang Technological University, Singapore

W **World Scientific**

NEW JERSEY · LONDON · SINGAPORE · BEIJING · SHANGHAI · HONG KONG · TAIPEI · CHENNAI · TOKYO

Published by

World Scientific Publishing Co. Pte. Ltd.

5 Toh Tuck Link, Singapore 596224

USA office: 27 Warren Street, Suite 401-402, Hackensack, NJ 07601

UK office: 57 Shelton Street, Covent Garden, London WC2H 9HE

Library of Congress Control Number: 2021048500

British Library Cataloguing-in-Publication Data
A catalogue record for this book is available from the British Library.

STRATEGY AND DEFENSE POLICY FOR SMALL STATES
Problems and Prospects

ISBN 978-981-124-428-5 (hardcover)
ISBN 978-981-124-429-2 (ebook for institutions)
ISBN 978-981-124-430-8 (ebook for individuals)

For any available supplementary material, please visit
https://www.worldscientific.com/worldscibooks/10.1142/12484#t=suppl

Desk Editors: Balamurugan Rajendran/Thaheera Althaf

Typeset by Stallion Press
Email: enquiries@stallionpress.com

Preface

When Dr Tony Tan returned to Cabinet in 1995 as Deputy Prime Minister and Minister for Defence, the world was still in the midst of momentous strategic changes. The Soviet Union had collapsed and the Cold War had consequently ended, and there were proclamations made, particularly in Europe but elsewhere too, that a "peace dividend" would come about, and military organisations in North America and western Europe began the process of downsizing. There were questions as to whether or not the United States would remain engaged and committed to the security and stability of the Asia Pacific region. Furthermore, because of the downsizing of United States and western European military organizations, the global arms market was being flooded with modern military technologies and weapons systems, because the defense industries of these countries now had excess stocks and production capacity. Underpinned by strong economic growth, military organizations elsewhere began rapid modernization of their existing orders of battle.

At the same time, the ending of the Cold War had removed some of the strategic restraints in many parts of the world that the United States and the former Soviet Union had imposed to ensure that nuclear Armageddon did not happen. Freed from these restraints, many so-called small wars began to break out: in Rwanda, South Ossetia in the former Soviet republic of Georgia, and in the former Yugoslavia, among others. Furthermore terrorism was an on-going security problem for many states: the Aum Shinrikyo sarin gas attack in Tokyo on 20 March 1995 and the Oklahoma City bombing on 19 April 1995, among others. From the perspective of a small state like Singapore, it seemed the end of the Cold War

v

did not provide any peace dividend; instead, it had ushered in a period of great uncertainty in the Asia Pacific.

It was in this context that Dr Tan decided that the Ministry of Defence would benefit from having an additional source of geopolitical analysis that could help refine Singapore's defense and foreign policies, as the country sought to navigate its way through the geopolitical and geostrategic uncertainty in the Asia Pacific region arising from the post-Cold War era. Dr Tan subsequently established the Institute of Defence and Strategic Studies (IDSS) in 1996 as a think tank that would conduct research on the geopolitical and geostrategic developments in the Asia Pacific, ask the difficult—even inconvenient—questions that could have a deleterious impact on Singapore's national security, and where necessary and possible provide alternative policy thinking. Mr SR Nathan, then Ambassador to the United States, was tapped to be the founding Director of IDSS.

Building on its research agenda, IDSS would have two inter-related additional functions. One, IDSS would assume a graduate education function, in the form of a Master of Science degree programme in Strategic Studies established in 1998. In 2002 and 2004, two new degree programmes—on International Relations and International Political Economy respectively—were added to the education function. In 2008, after the S Rajaratnam School of International Studies (RSIS) was established, and IDSS became a constituent institute of RSIS, a new degree programme in Asian Studies was added. To express the growth in other ways: in 1998, there were a total of 8 enrolled students, all Singaporeans; today, the student body numbers over 200 students in the four aforementioned degree programmes including a PhD programme, coming to Singapore from throughout Asia, Africa, Europe, and North and South America.

In addition to its own MSc degree programmes, IDSS became the main provider of Professional Military Education for the officer corps of the Singapore Armed Forces. Beginning in 1998, IDSS began to deliver lessons on military history, regional geopolitical issues and strategic concepts to the Singapore Armed Forces officer corps at all levels of officer education, in particular, at the Singapore Command and Staff College. In 2008, this responsibility for the major component of the SAF officer corps' PME was enhanced, with the establishment of a formal partnership between the SAF and Nanyang Technological University, involving RSIS—which had already taken over the mantle of IDSS—and the Nanyang Business School and the College of Engineering, providing for

MSc-level PME at the former Singapore Command and Staff College, today renamed as the Goh Keng Swee Command and Staff College (GKSCSC) in honor of Singapore's first Minister for Defence. These courses allow graduates of the GKSCSC to complete graduate degrees at NTU. This commitment to the SAF's PME underscores the raison d'etre of IDSS—that the increasingly complex security issues that Singapore faces will require an increasingly well-educated strategic community able to understand the geopolitical and geostrategic context in which Singapore finds itself, and thereafter navigate the country through the issues and problems that it will face.

Two, IDSS would test its research output in the "marketplace" of ideas, through networking with other like-minded research institutes and think tanks. IDSS created a number of forums that served as platforms for strategic thinkers and policy makers to present their arguments about strategic issues facing the Asia Pacific. The Strategic Challenges in the Asia Pacific Distinguished Lecture series featured policy makers such as the United States Secretary of Defence as well as the Minister for Defence for the People's Republic of China. A Summer School for senior military officers—the Asia Pacific Programme for Senior Military Officers—created a platform through which senior military officers from the Asia Pacific region and beyond could come together for an intensive week-long programme of talks and exchanges by leading academic thinkers and policy makers. IDSS also became the Singapore National Committee of the Council for Security Cooperation in the Asia Pacific (CSCAP), a network of security-focused think tanks and institutions that advises their respective government representatives in the ASEAN Regional Forum.

IDSS has not remained static. As new areas that can impinge on Singapore's national security emerge and develop, IDSS has had to adapt and enlarge, where necessary. With the emergence of a new terrorist threat to Singapore—bookmarked by the 11 September 2001 terrorist attacks, and further accentuated since then by other terrorist attacks in Bali, London and Paris, amongst others—the International Centre for Political Violence and Terrorism Research (ICPVTR) was formally inaugurated on 20 February 2004. Subsequently, new research programmes were added—the Maritime Security Programme and the Military Transformations Programme. At the same time, older research programmes have continued to refine and adjust their respective areas of focus, in step with the ever-evolving strategic landscape that Singapore finds itself in. However, as IDSS has evolved, one aspect has remained constant: its desire to connect academically rigorous thinking to Singapore's national security requirements. It is in that spirit that this collection of essays was conceived. The essays revolve around themes and concepts derived from the academic study of strategy and security, and seek to apply them to the strategic considerations of small states such as Singapore.

About the Editor

Bernard F.W. Loo (isfwloo@ntu.edu.sg) is Senior Fellow with the Military Studies Programme and concurrently Coordinator, Master of Science (Strategic Studies) degree programme at the S. Rajaratnam School of International Studies (RSIS), Nanyang Technological University, Singapore. He is the author of *Medium Powers and Accidental Wars: A study in Conventional Strategic Stability* (Edwin Mellen, 2005), and the editor of *Military Transformation and Operations* (Routledge, 2009). The latter title was translated into complex Chinese. His other publications have appeared in the *Journal of Strategic Studies, Contemporary Southeast Asia, NIDS Security Reports, Prism,* and *Taiwan Defense Affairs.* He has spoken on defense and strategic issues in institutions and conferences in China, Estonia, Finland, Japan, Latvia, Lithuania, New Zealand, the Philippines, South Korea, Taiwan, and the United States.

About the Contributors

Alistair D. B. Cook (iscook@ntu.edu.sg) is Coordinator of the Humanitarian Assistance and Disaster Relief Programme and Senior Fellow at the Centre for Non-Traditional Security Studies (NTS Centre), S. Rajaratnam School of International Studies (RSIS), Nanyang Technological University (NTU) Singapore. His research interests focus geographically on the Asia-Pacific and Myanmar in particular and thematically on humanitarian assistance and disaster relief (HADR), foreign policy and regional cooperation. He has taught undergraduate, graduate and professional development courses at Purdue University, University of Melbourne, Deakin University, Nanyang Technological University, Australian National University, Singapore Civil Defence Academy and SAFTI.

Weichong Ong (iswcong@ntu.edu.sg) is an Assistant Professor with the Military Studies Programme, the Institute of Defence and Strategic Studies (IDSS), the S. Rajaratnam School of International Studies (RSIS), Nanyang Technological University, Singapore. He is concurrently is an Affiliated Researcher with the Department of Leadership and Management, Swedish National Defence College and a Guest Professor at the Ecole Navale, France. He is also an alumnus of the Phillip Merrill Center for Strategic Studies Basin Harbor Workshop, SAIS, Johns Hopkins (2013). Weichong is the author of *Malaysia's Defeat of Armed Communism: The Second Emergency, 1968–1989* (Routledge, 2014). In addition, he has published articles in the *Australian Defence Force Journal*, *Armed Forces and Society*, *Asia Policy*, *Canadian Naval Review*, *Defence Studies*,

Journal of Southeast Asian Studies, *The Pacific Review*, and the *RUSI Journal*. His op-eds and commentaries have appeared in *Diplomatie*, *The National Interest*, *Defense News*, *The Diplomat*, *The Straits Times*, *Today*, and the *Lianhe Zaobao*.

Hikaru Yamashita (hikaru@u-shizuoka-zen.ac.jp) is Professor in the Department of International Relations at the University of Shizuoka, Japan. Prior to joining the University in 2020 he worked for the National Institute for Defense Studies for 17 years. He is the author of *Humanitarian Space and International Politics: The Creation of Safe Areas* (2004), *Evolving Patterns of Peacekeeping: International Cooperation at Work* (2017) as well as journal articles and commentaries on peace operations, global governance, and international theory. He was a Fulbright Visiting Scholar at Columbia University's Saltzman Institute of War and Peace Studies between 2008 and 2009.

Acknowledgments

This project was conceptualized as part of a series of events to commemorate the 25th anniversary of the Institute of Defence and Strategic Studies (IDSS). When I was tasked with managing this, I had thought of a volume of essays that would reflect the wide-ranging work of IDSS. Unfortunately, the twin challenges of Covid-19 and personal work meant that many colleagues were unable to contribute. I am very grateful, therefore, that there were nevertheless friends and colleagues near and far who did answer my pleas: Dr Alistair Cook, Professor Hikaru Yamashita, and Assistant Professor Ong Wei Chong. This volume of essays would not have been possible without their invaluable contributions.

My gratitude also goes to my colleague, Mrs Ong-Chew Peck Wan, who was extremely helpful in identifying potential publishers, and whose constant reminders ensured that this volume would not suffer from my tendency to procrastinate. That this was completed in time is really down to her!!!

My thanks, also, to the staff at World Scientific, who have worked tirelessly to get this completed in time, despite the short time frame, and who designed the cover art.

Weichong Ong's chapter, "Singapore's Military History 2.0," is an expanded version of "Singapore's New Military History: A Military History from a Non-War Fighting Past," in Tristan Moss and Rom Richardson (eds.), *Beyond Combat: Australian Military Activity Away From the Battlefields* (Sydney: UNSW Press, 2018). I would like to express my gratitude to UNSW Press for their kind permission to have this version of Ong's earlier chapter reproduced here.

This collection of essays is dedicated to the memory of the late Mr SR Nathan, the founding Director of IDSS and the 6th President of the Republic of Singapore. IDSS today is a reflection of his thoughtful leadership.

Contents

Preface   v

About the Editor   ix

About the Contributors   xi

Acknowledgments   xiii

Introduction   *Towards a Strategic Studies for Small States*   xvii
                 Bernard F.W. Loo

Chapter 1   Geopolitics, Geostrategy, and Defense Policy   1
            Bernard F.W. Loo

Chapter 2   Strategic Culture and Strategy   19
            Bernard F.W. Loo

Chapter 3   Singapore's Military History 2.0   41
            Weichong Ong

Chapter 4   Military Modernization in the 21st Century Problems
            and Prospects for Small Military Organizations   57
            Bernard F.W. Loo

Chapter 5   The Changing Character of War in the 21st Century
            Challenges for Strategic Planning   81
            Bernard F.W. Loo

Chapter 6   Military Engagement in Disaster Response Policies,
            Interests and Issues   105
            Alistair D.B. Cook

Chapter 7   Peace Operations and Military Organizations
            Between Internationalism and Statism   131
            Hikaru Yamashita

Index   151

https://doi.org/10.1142/9789811244292_0001

Introduction

Towards a Strategic Studies for Small States

Bernard F.W. Loo

Senior Fellow, Institute of Defence and Strategic Studies,
S Rajaratnam School of Internal Studies,
Nanyang Technological University, Singapore

isfwloo@ntu.edu.sg

In 1996, then-Deputy Prime Minister and Minister for Defence of the Republic of Singapore, Dr Tony Tan established the Institute of Defence and Strategic Studies (IDSS), with the express purpose of conducting "research on geopolitical developments in the Asia Pacific region and to promote greater understanding of, and interest in, strategic and defense issues in Singapore and the region." The IDSS mission was to "inform policy makers of the political uncertainty in the Asia Pacific region arising from the post-Cold War era and the possible implications of this uncertainty on Singapore's security and defense." From its inception, the IDSS would also offer a Master of Science degree in Strategic Studies, the student body of which coming principally, if not exclusively, from the Singapore Ministry of Defence and the Singapore Armed Forces.

If the IDSS was formed with geopolitical developments as the core focus of its research, it reflects a particular understanding of what Strategic Studies encompasses; indeed, it could be argued that the focus on geopolitics set the IDSS apart from its peer institutions. For instance, the Philip Merrill Center for Strategic Studies at the School of Advanced International Studies, Johns Hopkins University "operates as a center for strategic thought and education [and] explores the relationship between politics and the many kinds of military power—from the use of terror by small, non-state groups, to the threatened use of nuclear weapons—aiming to promote dialogue and innovative research on pressing national security issues."[1] The Strategic and Defence Studies Centre at the Australian National University focuses on "the analysis of the use of armed force in its political context [and the] primary expertise within the broad field of Strategic Studies consists of three related research clusters: Australian defense, military studies, and Asia-Pacific security. understanding the complexity of Asia's strategic environment, Australia's place in it, and the utility and application of armed force in international affairs."[2] The International Institute of Strategic Studies describes itself as a "research institute that provides objective information on military, geopolitical and geo-economic developments that could lead to conflict."[3]

It is obvious that how these different institutions have conceptualized Strategic Studies is subtly, but importantly, different. Furthermore, how the subject is taught is similarly different. At the S Rajaratnam School of International Studies, to which the IDSS belongs, the Master of Science degree in Strategic Studies focuses on the "study of the preparation, threat, use and control of organized force, by both state and non-state actors, for political purposes." At the University of Aberdeen, its Master of Science in Strategic Studies[4] is described as "multidisciplinary and international in course content." Its focus is "real world security issues which challenge governments, conflict, armed services, peace, international organizations and business around the world," and students will be equipped with "advanced theoretical and operational understanding of the characteristics, practice and profound effects of use of force by states and non-state actors throughout the international system." The Victoria University of Wellington in New Zealand[5] offers postgraduate qualifications in Strategic Studies, where its students will develop "a deep practical and theoretical understanding of the ways power is exercised ... how conflict is managed and security maintained across the globe." Students will "get insights into the pressing geopolitical issues ... such as the relationship between the US and China, cybersecurity, the South China Sea, and strategic responses to the problems of terrorism and civil conflict."

Within Southeast Asia, the Association of Southeast Asian Nations' Institutes of Strategic and International Studies (ASEAN-ISIS) network of think tanks and research institutes comprises the following member organizations: Brunei Darussalam Institute of Policy and Strategic Studies; the Cambodian Institute for Cooperation and Peace; the Centre for Strategic and International Studies (Indonesia); the Institute of Foreign Affairs (Lao PDR); Institute of Strategic and International Studies (Malaysia); Myanmar Institute of Strategic and International Studies and International Studies; the Asia Pacific Pathways to Progress Foundation. (The Philippines); the Singapore Institute of International Affairs; the Institute of Security and International Studies (Thailand); and the Diplomatic Academy of Vietnam. If the four institutions mentioned earlier reflect the scope of Strategic Studies, it is worth noting that defense and strategic issues and challenges are not necessarily universally addressed by the 10 member organizations of ASEAN-ISIS.

This, then, begs the question: what is the Strategic Studies that the IDSS was established to conduct research into and teach? That is the purpose of this collection of essays: to lay out how Strategic Studies is understood at the IDSS, and how the subject is taught at the IDSS's parent institution, the S Rajaratnam School of International Studies. Its starting point is the proposition that the IDSS has a particular vision of what Strategic Studies is, especially in terms of how the subject is taught. This vision of the subject is a function of the geopolitical context in which the IDSS exists, being part of a graduate school of international studies located in the Republic of Singapore. This proposition therefore echoes, and seeks to address, the argument of Amitav Acharya and Jiejie He: that "Western writings on strategy, on such supposedly universal concepts of war or peace, draw heavily from the ideas, agency, and historical experiences of Europe [and] often neglect or give only token space to other historical experiences and cases."[6] It is predicated on the Clausewitzian idea that if war has a logic and grammar—where its logic remains constant, but its grammar is ever-changing—this logic-grammar relationship exists in strategy as well. The idea that strategy has a logic that is universal and grammar that is ever-changing echoes the argument made by Barry Buzan in 1987: "Although the surface subject matter of Strategic Studies changes quickly as new technologies and new conflicts come and go, its concepts are relatively stable. Ideas like disarmament and arms racing are very longstanding, and even more recent additions to the strategic vocabulary such as deterrence and arms control have been central to strategic discourse for several decades."[7] Nevertheless, while strategy may have a logic that is universal—the logic that underpins strategy for Great Powers is not fundamentally different from that which underpins strategy for small states—the grammar of strategy changes from actor to actor, reflecting the unique geographic, cultural, and material conditions of each actor. In this regard, this study agrees in part with the assessment of Acharya and He: that "supposedly universal strategic concepts can have different meanings and applications in different contexts."[8]

Granted, Clausewitz continues to be a somewhat divisive character in the world of Strategic Studies, and his works, and in particular *On War*, continue to inspire an essentially binary reaction from readers from the academic and policy-making worlds. Studies on Clausewitz—both positive and negative—have remained something of a cottage industry, nearly two hundred years after his death from cholera in 1831. Indeed, beginning with the British strategist, Basil Henry Liddell Hart, a remarkably resilient cottage industry of Clausewitz-bashing—to borrow from the words of Christopher Bassford, something of a "grand tradition of trashing Clausewitz."[9] Recently there has been a spate of publications that seeks to reinforce—if not re-establish—Clausewitz's legacy as the single most important strategic thinker.[10]

This study is not the place to discuss the merits or otherwise of Clausewitz's paradigm of strategy and war. Suffice it to say that the authors of this study consider Colin Gray's assessment of Clausewitz to be essentially correct: that Clausewitz offered "the most persuasive of intellectual road-maps to the nature and function of strategy and war"; even if the "weaknesses in his theory … stemmed principally from his place in time and culture," it has left "considerable terrain untrod and he was not infallible," and "his ideas [requiring] periodic amendment and clarification."[11]

To return, therefore, to the logic-grammar argument, this volume of essays is an attempt to lay out the Strategic Studies that is reflected in the research output of the IDSS and in the Master of Science in Strategic Studies degree taught at the S Rajaratnam School of International Studies. The logic of Strategic Studies may be universal—concepts such as arms dynamics and military modernization, arms racing and the security dilemma, deterrence, offence-defense theory surely apply to the strategies and defense policies of all states—but the grammar, that is, how these concepts are translated into actual defense policy and strategy will differ from state to state, as geographical, historical, cultural, technological and other material conditions change from one state to another. This issue will be discussed in greater depth later.

Defining Strategic Studies

What, then, is Strategic Studies? In a sense, the name of the subject offers its own answer: the study of strategy; however, it is obvious that this answer is tautological, and therefore, unsatisfactory. Furthermore, what constitutes 'strategy' varies, depending on whom the question is posed. Given the Clausewitzian logic that frames this study, perhaps it is best to begin with how Clausewitz defined strategy.

To Clausewitz, strategy is "the use of engagements for the object of the war."[12] In this one phrase, it can appear that strategy necessarily resides in the phenomenon of war; war is explicitly mentioned, and the term "engagements" can be interpreted as a reference to battles. In *The Direction of War*, Hew Strachan further develops this idea: "Strategy is a profoundly pragmatic business: it is about doing things, about applying means to ends. It is an attempt to make concrete a set of objectives through the application of military force to a particular case … In other words, strategy lies at the interface between operational capabilities and political objectives; it is the glue which binds each to the other and gives both sense … it is based on a recognition of the nature of war itself."[13] In other words, strategy is necessarily connected to the phenomenon of war, of the application of military power—violence, actually—in the context of mutually hostile political actors, each seeking to impose its agenda over the other.

However, other scholars sought to expand the focus of strategy. John Garnett was one such scholar, as shown in his introductory essay to *Contemporary Strategy I: Theories and Concepts*, first published in 1975.[14] In it, Garnett argued, "For the man in the street, strategy is intimately connected with planning wars and fighting them. It is a military activity par excellence in which high-ranking officers plan the overall conduct of wars … But in a very important sense the man in the street has got it wrong." Rather, Garnett argued that 'strategy' ought to be understood as "the ways in which military power may be used to achieve political objectives," and that war is merely "one of the ways in which military power can be used to implement political goals." Garnett went so far as to subsequently assert, "Today, purely military definitions of strategy have virtually disappeared because they have failed to convey either the flavor or the scope of a subject that straddles the spectrum of war and peace and is as much concerned with statesmanship as with generalship."

The argument that strategy encompasses more than just war fighting has since been increasingly accepted. In 1987, citing Louis Halle's *The Elements of International Strategy*, Barry Buzan defined Strategic Studies as "the branch of political studies concerned with the political implications of the war-making capacity available to nations ... the use of force in political relations within and between states ... the instruments of force, and the way in which those instruments affect relations among the states that possess them."[15] More recent definitions of strategy may begin to fudge the centrality of combat, for instance, Colin Gray's rather elegant definition of strategy as the "bridge that relates military power to political purpose."[16] Arguably, Gray might still imply the centrality of battle—after all, military power without being applied in warfare is potential rather than actual—but his definition does not necessarily preclude the use of threats, underpinned by the potential power of the state's military instruments, of course. The inclusion of threats, in other words, does not negate the central role that combat plays in conceptions of strategy.

Strategic Studies is understood by its scholars as a multi-disciplinary subject area, drawing on the humanities (history, and especially military history), the social sciences (anthropology, economics, political science and international studies, psychology and sociology), and the hard sciences (technology, in particular). Its focus, strategy, is, in the words of Hew Strachan, "a profoundly pragmatic business; it is about doing things about applying means to ends."[17] Strategic Studies traditionally situates itself within the broader ambit of international studies, and locates itself in the realm of modern states; the underlying premise is that only states can have the legitimate right to use force in pursuit of their political objectives. However, given the prevalence of intra-state or civil wars, and the phenomenon of international terrorism, the subject can expand its focus to include non-state actors such as terrorist groups, insurgency movements and rebel groups; in so doing, the arena of Strategic Studies can be enlarged to include politics within states. That being said, Strategic Studies has evolved across time and space.

Strategic Studies, as recounted by John Baylis and James Wirtz, owes its emergence to the efforts of the American strategist Bernard Brodie, who argued that the study of strategy required "a more rigorous, systematic for of analysis of strategic issues compared with the rather narrow approach to security problems adopted by the military, who were preoccupied with tactics and technology." However, by the 1960s, Brodie came to a different conclusion, that the Strategic Studies he helped found had developed a potentially fatal flaw: the emphasis on scientific rigor had turned into a "scientistic strain" that resulted in an "'astonishing lack of political sense' and the 'ignorance of diplomatic and military history' that seemed to be evident among those writing about strategy."[18] As a result, what Strategic Studies actually focuses on changes over time. In the 1950s, Strategic Studies was very much focused on the central issue of the Cold War: "how to survive and prosper in the nuclear age, when Armageddon might be just minutes away."[19] The scholars, strategic planners and policy makers of the time, having lived through World War Two—and witnessed how the preceding utopianism in international politics had failed to prevent it from happening—came to the conclusion that the world was inherently anarchic in nature. Consequently, Strategic Studies traditionally has been predicated on an essentially Realist conception of international politics; as Buzan described it, "composed of independent political entities ... mostly states [who] all possess the capability to use force against each other to some degree, and their interests conflict with sufficient frequency and intensity that the threat of force is an unavoidable and constant feature of their existence."[20] The "relatively peaceful collapse of the Soviet Union" did not necessarily end the Cold War's concerns about potential nuclear Armageddon; nevertheless, it heralded the return of utopian conceptions of international politics, and the increasing suspicion that Strategic Studies, with Realism as its core assumption, was "part of the problem of international security," and its scholars as "unwilling to come to terms with the fact that force was apparently fading as a factor in world politics."[21]

There is also a spatial aspect to the focus of Strategic Studies. At least in the English medium literature, there appears to be an oceanic divide between two broad approaches to Strategic Studies, and this is encapsulated in the coverage of two of the subject's most important journals, *International Security* and *The Journal of Strategic Studies*. The former directs its attention largely, although not exclusively, on matters pertaining to current strategic problems, with a very strong policy-relevant angle to its publications. During the height of the Cold War, it was not uncommon to find articles in *International Security* that focused on problems of nuclear deterrence, problems of conventional military balances between the Warsaw Pact and NATO. Recent issues of the journal have looked at, among other things, the emergence of global terrorism as the new security preoccupation of states. In much of this literature, there has always been a strong element of policy relevance, of even policy advocacy. The latter tends towards a more historical bias, many of its publications falling within the broader ambit of military history and military theory and thought. Many of its articles have addressed problems of military history, either in the form of campaign or war studies on the one hand, or of military thought on the other. It is therefore not uncommon to find articles examining the strategic thought of thinkers such as Basil Henry Liddell Hart, or articles re-examining the military history of World War Two in this journal. The journal would also publish articles on political and military geography, on the evolution of armed forces. Its direct policy relevance might not always be immediately evident. As much as Strategic Studies is a policy-relevant subject, it is therefore possible to see it in purely academic terms (albeit not without its policy relevance).

Towards a Strategic Studies for Small States

This collection of essays seeks to address three sets of issues that Strategic Studies has had to grapple with. To begin with, and returning to the argument by Acharya and He mentioned earlier, much of Strategic Studies is Western in its origins, assumptions, and focus. As stated earlier, while this study accepts the proposition that strategy, like war, as a logic that is universal, its grammar—that is, how strategy translates from an intellectual into a pragmatic exercise—ought to reflect the unique geopolitical and cultural-historical conditions of its practitioners. The purpose of this collection of essays is to begin to move towards developing a Strategic Studies that may be underpinned by a universal logic, but is written nonetheless to address the conditions that small states face. Furthermore, given its focus during the Cold War, it is possible to identify a second set of issues in Strategic Studies, namely, conventional strategy, which had not been studied with the same rigor as nuclear strategy. As noted by Joseph Nye and Sean Lynn-Jones in 1988, while Strategic Studies had justifiably focused its attention on the challenge of managing the strategic rivalry between the superpowers, conventional strategy also deserved serious study: since 1945, all wars fought have been at the conventional level, and even a superpower nuclear war would have likely begun at the conventional level.[22] Finally, given the definition of Strategic Studies by RSIS, the use of "organized force" by "both state and non-state actors, for political purposes" means that, for many small states, military organizations are deployed for a range of operations and missions that fall outside of the phenomenon of war that General Sir Rupert Smith describes in his 2005 book, *The Utility of Force*: "battle in a field between men and machinery, war as a massive deciding event in a dispute in international affairs."[23] Military organizations have had to plan, and even write doctrine, for the employment of military forces in a range of so-called Operations Other Than War. These types of missions therefore similarly deserve serious attention in Strategic Studies.

One of the first attempts at addressing the logic of strategy was published in 1975. *Contemporary Strategy I: Theories and Concepts*, edited by John Baylis, Ken Booth, John Garnett and Phil Williams, discussed the concepts that underpinned Strategic Studies. Its contents covered: the evolution of strategic thinking; the role of military power; the relationship between technology and strategy; nuclear deterrence; arms control and disarmament; theories of limited war, revolutionary war; crisis management; and alliances. A second volume, edited by the same august scholars, focused explicitly and solely on the nuclear powers. However, even the first volume, which was purportedly conceptual in its focus, had a very specific location in its temporal and geopolitical focus: the Cold War, and the problems of strategy facing the major powers.

Buzan's 1987 book is important, because it represents one of the first major attempts at addressing the logic of Strategic Studies. As Buzan argued,

> The need for an introduction to Strategic Studies stems most obviously from the size of the strategic literature. More than three decades of writing have accumulated since the first competitive deployments of nuclear weapons caused Strategic Studies to emerge as a distinct field during the 1950s ... During the last 30 years the expansion of strategic literature has been driven by fast-moving developments in technology, conflict and politics. These range from new weapons, like cruise missiles, to new wars, like that in the Gulf, to changes in political alignments, like the Sino-Soviet split ... [the literature] mostly reflects an intense, short-term policy orientation that is closely tied to the agenda of government decision-making on defence and military issues ... It suits experts, because it enables them to structure their own writing easily ... It is, however, a

barrier to those seeking entry to the subject. It confronts them with an unassembled jigsaw puzzle of parts with little guide as to how they all fit together ... The structure of strategic literature does not clearly reveal the essentials of the subject, and therefore does not serve the needs of the many non-experts who rightly feel that they want to understand what is going on.[24]

However, is it correct to make a distinction between nuclear and conventional strategy, in the first place? Does strategy really have a grammar that changes constantly, if its logic is truly constant and universal? Colin Gray had consistently argued that there is no distinction between nuclear and conventional strategy. In 1982, he stated, "the political world had not altered strategic questions or rendered them any less salient than they were prior to Nagasaki and Hiroshima."[25] In 1988, he argued that all nuclear weapons did was to make policy-makers more cautious in their inter-state dealings with each other.[26] Any attempts at constructing a distinction between nuclear and conventional—manifested, for instance, in John Mearsheimer's *Conventional Deterrence*—were therefore inherently flawed.[27] Or, as Bradley Klein's rhetorical question posed, "How else to make sense of the claim, so widespread after Hiroshima, of a 'nuclear revolution', as if the concept of a revolution could be explained in terms of a technological change in weapons systems."[28] However, this study argues that differences emerge in how the universal logic is translated or manifested, in the way policy-makers, scholars and strategists approach the application and implementation of strategy from state to state. If nothing else, differences in the range of strategic instruments available to individual states means the strategic choices that their policy makers adopt will reflect these differences in available instruments and resources. Nothing illustrates the differences between the nuclear and conventional dimensions of strategy more clearly than deterrence postures.

Two widely known statements about deterrence capture both the universal logic and the variegated grammar of conventional versus nuclear strategy. The first statement, forever associated with nuclear deterrence, comes from Bernard Brodie's 1946 book, *The Absolute Weapon*: "Thus far the chief purpose of our military establishment has been to win wars. From now on its chief purpose must be to avert them. It can have almost no other useful purpose." At least for the United States, nuclear deterrence came to be understood in essentially counter-value, or punishment, terms: it was the idea that the former Soviet Union could launch a massive first strike with its nuclear weapons, and the United States would still have enough nuclear weapons to launch a retaliatory attack that would inflict unacceptable levels of damage on the urban and industrial centers of its Cold War adversary. The second statement comes from the 4th Century Roman strategist, Publius Flavius Vegetius Renatus: "Si vis pacem, para-bellum (If you want peace, you must prepare for war)." Conventional deterrence has to reflect the technological limitations of non-nuclear weapons; given the chasm in destructive power between nuclear and conventional weapons, it is therefore difficult to see how a deterrence posture based on punishment of initial aggression could apply to conventionally armed states. Rather, as John Mearsheimer argued, conventional deterrence would have to be based on the ability to persuade a would-be aggressor that its plans for war are unlikely to succeed, and hence, the actual resort to war would be strategically counter-productive.[29]

The two manifestations of deterrence theory therefore betray a fundamentally different attitude towards the strategic prospect of war. If nuclear deterrence postulated the absolute need to avoid war, conventional deterrence has traditionally been predicated on war-fighting; indeed, as the Vegetius' statement mentioned above indicates, the ability to wage war (and presumably successfully, in other words be able to deny the aggressor any prospect of strategic success) is central to conventional deterrence. Brodie's 1946 observation is, arguably, correct, that the actual application of nuclear weapons in a superpower war was not only counter-productive, but also inherently contradictory: no *raison d'etat* could possibly justify a nuclear holocaust, despite the existence of arguments supporting the notion of nuclear warfighting.[30] Thus, while the nature of strategy may not have changed, the scope of strategic actions available to nuclear states had. Strategy, at least for conventional states, does not exclude the possibility that the military instruments will not be used in war. Policy makers, scholars and strategists have rather less contested the notion that conventional states reserved the right to resort to war as and when necessary. For nuclear states, strategy was now about the prevention of nuclear war; while conventional states, however, have continued to wage war when necessary.

Finally, while military organizations exist to protect their respective states from armed aggression, the operational reality for many military organizations around the world is that they have not waged war against other military organizations, but they have deployed for so-called operations other than war (OOTW)—including disaster relief operations, a spectrum of peace operations and counter-terrorism operations. The inclusion of OOTW into the missions that military organizations have to prepare for reflects in part the expansion of the meaning of security. Strategic Studies has traditionally understood the state to be the referent object of security. However, the concept of security has been steadily expanding, to incorporate a number of new definitions and aspects.[31] Security is now thought of not simply in terms of the state, but also in terms of individual human beings as an equally important referent of the concept. If military organizations are deployed to mount disaster relief operations, or conduct peace operations under the banner of the United Nations Department of Peacekeeping Operations, or contribute significant resources to work in tandem with law enforcement agencies in counter-terrorism operations; and if Strategic Studies at RSIS is the "study of the preparation, threat, use and control of organized force, by both state and non-state actors, for political purposes"; how these military forces are employed, in support of the broader political interests of the state, are worthy of serious study within the ambit of Strategic Studies.

This collection of essays focuses on how small states can think about the preparation for the use of force. It looks at ways through which small states can think about the issues that underpin thinking about defense policies, understood here as how a state's policy makers and strategic planners think about the state can protect it from the threats the state faces, whether real or imagined. It is organized around two themes.

The first theme focuses on a number of key concepts in Strategic Studies, in the context of small states. These concepts help to explain how and why the state's policy makers and strategic planners come to identify the security concerns and threats that the state faces. To develop this theme, Chapter One discusses the relationship between geography and defense policy. It focuses on the concepts of geopolitics and geostrategy, and how these concepts set the context, both physical as well as psychological, in which policy makers and strategic planners of small states operate; the perceptions of the strategic environment in which the state exists; and how these actors begin to understand the potential threats and security challenges that the state is likely to face. Chapter Two further develops this theme, by turning its attention to the concept of strategic culture. In the political and strategic geographic context of small states, when paired with a less than positive military history, this can generate for small states particular tendencies and preferences in terms of the strategic scenarios and options that military planners are likely to provide their policy makers, and the eventual strategic choices of these policy makers. In Chapter Three, Weichong Ong examines, in a case study of Singapore, how military history can be used to provide another basis for strategic choices of a state. Given that many states are the creation of the decolonization processes that were unleashed after World War II, policy makers can, and in the case of Singapore do, use the military history of their colonial past as the historical basis for the state's strategic culture.

The second theme centers on contemporary challenges and dilemmas that ultimately question the continuing relevance of the so-called traditional paradigm of Strategic Studies. Chapter Five focuses on the questions that small states have to address in engaging in what Barry Buzan called the arms dynamic, a process by which military instruments are first constructed, subsequently modernized as and when necessary and feasible, and even redesigned to meet emerging security concerns. Presumably, states will design their respective military organizations to be able to address the security challenges and threats they are considered likely to face. Chapter Six furthers this theme by examining the character of war in the 21st Century. It posits that the character of war in the 21st Century will likely be significantly different from the character of wars—and especially inter-state wars—in the 20th Century. If this argument is correct, to revisit Chapter Six, states may be constructing a military instrument that is designed for wars of particular characteristics, and which will not be replicated in this century.

Then, the focus shifts to the operational challenges that many military organizations find themselves facing, with greater frequency, and these are operational challenges that are far from the worst case scenario of inter-state war. In Chapter Seven, Alistair Cooke examines the issue of disaster relief operations, and in Chapter Eight, Hikaru Yamashita discusses the issue of peace operations. In both disaster relief and peace operations, the respective authors argue that these are not operations that are part of the military organization's traditional skill-sets; nevertheless, given the strategic and diplomatic-political importance of such operations, military organizations will have to dedicate scarce time and financial resources to developing the material capacities, doctrines, mind-sets and skill-sets to successfully carry out such operations in the present and future. Both chapters therefore suggest that military organizations will face a dilemma: do they continue with "business as usual," that is, maintaining their focus on defending their respective states against armed threats, and therefore continue to acquire weapons systems and capabilities that allow them to prosecute such combat operations; or, do they "change course," and begin to re-orientate their organizations for non-combat operations, the so-called operations other than war?

Endnotes

[1] https://www.merrillcenter.sais-jhu.edu/about.

[2] http://sdsc.bellschool.anu.edu.au.

[3] https://www.iiss.org.

[4] https://www.abdn.ac.uk/study/postgraduate-taught/degree-programmes/309/strategic-studies/.

[5] https://www.wgtn.ac.nz/explore/postgraduate-programmes/master-of-strategic-studies/overview.

[6] Amitav Acharya and Jiajie He, "Strategic Studies: The West and the Rest," in in John Baylis, James J. Wirtz and Colin S. Gray (eds.), *Strategy in the Contemporary World: An Introduction to Strategic Studies* (Oxford: Oxford University Press, 2019), p. 330. Other scholars who have addressed the Western bias in Strategic Studies include, amongst others: Mohammed Ayoob, "Regional Security and the Third World," in Mohammed Ayoob (ed.), *Regional Security in the Third World: Case Studies from Southeast Asia and the Middle East* (London: Croom Helm, 1986); Edward Azar and Chung-in Moon, "Third World National Security: Toward a new conceptual framework," *International Interactions*, vol. 11, no. 2, 1984, pp. 103–135.

[7] Barry Buzan, *An Introduction to Strategic Studies: Military Technology and International Relations* (Houndsmill and London: The Macmillan Press, 1987), p. 8.

[8] Acharya and He, "Strategic Studies: The West and the Rest," p. 330. However, it needs to be pointed out that this study does not agree with the normative debates about the alleged ethnocentrism of Strategic Studies that Acharya and He sought to engage in. Acharya and He have a valid point in their insistence that there are regional and temporal manifestations to strategy; except they based their argument on a selective reading of Ken Booth's *Strategy and Ethnocentrism*, which was arguably less a critique of the subject of Strategic Studies as ethnocentric, and more an argument about the strategic perils of assuming the adversary to be a mirror image of oneself.

[9] Christopher Bassford, "John Keegan and the Grand Tradition of Trashing Clausewitz: A Polemic," *War in History*, vol. 1, no. 3, 1994, pp. 319–336. Indeed, Bassford characterizes Liddell Hart's Clausewitz-bashing as deliberate misrepresentation and nonsense.

[10] See, for instance: Antulio J. Echevarria II, *Clausewitz and Contemporary War* (Oxford Oxford University Press, 2007); Hew Strachan and Andreas Herberg-Rothe, *Clausewitz in the 21st Century* (Oxford: Oxford University Press, 2007); Christopher Coker, *Rebooting Clausewitz: 'On War' in the 21st Century* (Oxford: Oxford University Press, 2017).

[11] Colin S. Gray, *Modern Strategy* (Oxford: Oxford University Press, 1999), p. 359.

[12] Carl von Clausewitz (transl. Michael Howard and Peter Paret), *On War* (Princeton: New Jersey: Princeton University Press, 1984), p. 128.

[13] Hew Strachan, *The Direction of War: Contemporary Strategy in Historical Perspective* (Cambridge and New York: Cambridge University Press, 2013), p. 12.

[14] John Garnett, "Strategic Studies and its Assumptions," in John Baylis, Ken. Booth, John Garnett and Phil Williams (eds.), *Contemporary Strategy I: Theories and Concepts* 2nd edition (London and Sydney: Croom Helm, 1987), p. 3.

[15] Barry Buzan, *An Introduction to Strategic Studies: Military Technology and International Relations* (Houndsmill and London: The Macmillan Press, 1987), p. 4.

[16] Gray, *Modern Strategy*, p. 17.

[17] Strachan, *The Direction of War*, p. 12.

[18] John Baylis and James Wirtz, "Introduction: Strategy in the Contemporary World," in John Baylis, James J. Wirtz and Colin S. Gray (eds.), *Strategy in the Contemporary World: An Introduction to Strategic Studies* (Oxford: Oxford University Press, 2019), pp. 4–5.

[19] John Baylis and James Wirtz, "Introduction: Strategy in the Contemporary World," in John Baylis, James J. Wirtz and Colin S. Gray (eds.), *Strategy in the Contemporary World: An Introduction to Strategic Studies* (Oxford: Oxford University Press, 2019), p. 2.

[20] Buzan, *An Introduction to Strategic Studies*, p. 6.

[21] John Baylis and James Wirtz, "Introduction: Strategy in the Contemporary World," in John Baylis, James J. Wirtz and Colin S. Gray (eds.), *Strategy in the Contemporary World: An Introduction to Strategic Studies* (Oxford: Oxford University Press, 2019), p. 2.

[22] Joseph S. Nye Jr. and Sean M. Lynn-Jones, "International Security Studies: A Report of a Conference on the State of the Field," *International Security*, vol. 12, no. 4, Spring 1988, p. 24.

[23] Rupert Smith, *The Utility of Force* (London: Allen Lane, 2005), p. 3.

[24] Barry Buzan, *An Introduction to Strategic Studies: Military Technology and International Relations* (Houndsmill and London: The Macmillan Press, 1987), pp. 1–2.

[25] Colin S. Gray, *Strategic Studies: A Critical Assessment* (London: Aldwych Press, 1982), p. 7.

[26] Colin S. Gray, *The Geopolitics of Superpower* (Lexington: University Press of Kentucky, 1988), p. 30.

[27] Colin S. Gray, "Inescapable Geography," in Colin S. Gray and Geoffrey Sloan (eds.), *Geopolitics, Geography and Strategy* (London and Portland: Frank Cass, 1999), pp. 169–191. Also see: Gray, *Modern Strategy*, p. 322.

[28] Bradley S. Klein, *Strategic Studies and World Order: The Global Politics of Deterrence* (Cambridge, New York and Melbourne: Cambridge University Press, 1994), p. 2.

[29] John J. Mearsheimer, *Conventional Deterrence* (Ithaca, New York: Cornell University Press, 1983).

[30] The argument for nuclear warfighting was contested; see, for instance: Desmond Ball, "Toward a Critique of Strategic Nuclear Targeting," in Desmond Ball and Jeffrey Richelson (eds.), *Strategic Nuclear Targeting* (Ithaca and London: Cornell University Press, 1986), pp. 15–32; Desmond Ball and Robert C. Toth, "Revising the SIOP: Taking War-Fighting to Dangerous Extremes," *International Security*, vol. 14, no. 4, 1990, pp. 65–92; Harold A. Feiveson and John Duffield, "Stopping the Sea-Based Counterforce Threat," *International Security*, vol. 9, no. 1, 1984, pp. 187–202; Charles L. Glaser, "Why Even Good Defenses May Be Bad," *International Security*, vol. 9, no. 2, 1984, pp. 92–123; Jeffrey Richelson, "The Dilemmas of Counterpower Targeting," in Desmond Ball and Jeffrey Richelson (eds.), *Strategic Nuclear Targeting* (Ithaca and London: Cornell University Press, 1986), pp. 159–170.

[31] Barry Buzan, Ole Wæver, and Japp de Wilde, *Security: A New Framework For Analysis* (Boulder and London: Lynne Rienner Publishers, 1998).

https://doi.org/10.1142/9789811244292_0002

Chapter 1

Geopolitics, Geostrategy, and Defense Policy

Bernard F.W. Loo

Senior Fellow, Institute of Defence and Strategic Studies,
S Rajaratnam School of Internal Studies,
Nanyang Technological University, Singapore

isfwloo@ntu.edu.sg

This chapter posits the argument that geography is a necessary component of defense policy and strategy. If much of Strategic Studies tends to narrow its focus on the questions of war and peace to essentially military causes and to ignore the context in which particular inter-state relationships exist, this chapter will attempt to address that criticism, by examining the importance of the geographic milieu of defense policy and strategy.

The study of political geography (or geopolitics) highlights the importance of geography to international relations, as the study of strategic geography (or geostrategy) highlights the importance of geography to strategy and military operations. One author goes so far as to claim—with some degree of persuasiveness—that in the absence of a clearly identified enemy, geography can provide the basis for strategic planning.[1] Policymakers do not have to invent a potential adversary to be able to devise strategic plans; rather, they only need to analyze the geographic conditions that they are in, to determine its potential strategic weaknesses, and to devise the appropriate plans to counter them.

The literature on geopolitics and geostrategy is extensive.[2] This chapter will provide a review of the literature on geopolitics and geostrategy. Subsequently, this chapter will lay out the case for why these are necessary components to any study of defense policy and strategy.

Geopolitics

Political Geography, or Geopolitics, can be understood as the "spatial study and practice of international relations."[3] The subject draws its antecedents from the ideas of the key figures of the discipline, including Alfred Thayer Mahan, Halford Mackinder, Friedrich Ratzel, Rudolph Kjellen, and Nicholas Spykman. At the same time, Geopolitics has evolved intellectually, from a so-called Classical Geopolitics to a more recent trend of Critical Geopolitics. Indeed, for much of the 21st Century, Critical Geopolitics has, arguably, dominated the academic discussions, although in recent years, Classical (or, more likely a Neo-Classical) Geopolitics has enjoyed something of a resurgence.[4]

Classical Geopolitics is built on the premise that there are "certain geographical patterns in political history" that allow scholars and policy makers to derive "a theory of spatial relationships and historical causation."[5] Its focus is the relationship between politics and the physical environment, specifically how geography can provide an understanding of politics, in particular political decision-making.[6] This geography-politics relationship can be understood at three levels.[7]

First, geography is an object of policy. Territory is fundamental to statehood, which is, ultimately, a claim to have exclusive ownership and control over a particular piece of real estate.[8] A key focus of Classical Geopolitics is therefore the conflicts of interest between states, and how states may employ diplomatic, economic and military instruments of power to such conflicts.[9] The geography in question is presumably valuable in a tangible manner to the states involved: it may be materially valuable, for instance, possessing valuable raw materials, providing arable land for agriculture; or, it could have strategic value, such as allowing one state to escape potential strategic enclosure and subsequent strangulation or providing a buffer zone between otherwise hostile states. However, it is also possible that the geography in question is devoid of any material or strategic value, and be valued purely for its symbolism: state borders, as the subsequent discussion on Critical Geopolitics will demonstrate, maybe purely artificial, in most instances, but that does not negate the potential symbolism and consequent importance to the legitimacy of extant political elites.

Second, and following from the preceding observation about the necessary territoriality of statehood, geography provides the physical context of inter-state relations.[10] The China-United States relationship is necessarily maritime in character, as it exists in and is shaped by the Pacific Ocean that both connects and separates these superpowers. The historical relationship between the Korean peninsula and its Chinese and Japanese neighbors is very much shaped by the physical configuration—the manner in which the Korean peninsula juts out from the mainland, making the peninsula seem like "a dagger pointing at the heart of Japan and China."[11] The Singapore-Malaysia relationship has been, until recently shaped by the former's dependence on the latter for potable water, that dependency a function of the small size of Singapore and the previous lack of affordable technologies that would allow Singapore to escape this potential existential vulnerability.[12]

Third, should the geopolitical conflict of interest degenerate into war between the respective states, geography will matter in the theatre in which this war plays out.[13] A war between China and the United States, assuming it remains at the level of the non-nuclear, will necessarily be a maritime war, and will involve primarily, if not exclusively, the air and naval forces of both superpowers. Furthermore the fighting would take place over geography that is tactically or strategically significant to the outcome of such a war.[14] In World War II in the Pacific, the capture of Guam, Tinian and Wake islands was fundamentally important to the United States' counter offensive against Japan, because it was from these islands that long-range B-29 bombers could conduct the systematic destruction of Japanese cities through so-called strategic bombing. It was from Tinian island that the Enola Gay took off on its fateful mission over Hiroshima. This is an issue that will be further examined later in this chapter.

Classical Geopolitics may have been regarded as unfashionable in the aftermath of the end of the Cold War, in part because of its tendency to focus on geography as a source of potential conflict between states.[15] Ironically, this is precisely the attractiveness of Classical Geopolitics. As Benjamin Schreer has argued, in an era where a new Cold War seems imminent, this time between China and the United States, Classical Geopolitics provides an appropriate framework to study and understand the strategic behaviors of both China and the United States.[16]

As hinted above, Classical Geopolitics is not without its problems. Classical Geopolitics assumes the state to be a natural given, it emphasizes the geographical form of the state in absolute spatial terms, as well as the forces that promote or disturb territorial integration of differentiation. However, there is a case to be made that states are artificial constructs precisely because their borders are artificially constructed, in that they are the product of agreements between the policy makers of the states involved.[17] Granted, the state may be the result of a process of construction that, once constructed, makes the relationship between the ideational state and the territorial dimensions of the state appear to be almost inviolable. In other words, the state is the product of a naturalizing discourse that makes what is essentially social (and therefore non-permanent and changeable) seem natural (and therefore objective and eternal).[18] In treating the state as a natural given, Classical Geopolitics thus overemphasizes the impact of geography, and veer dangerously close towards geographical determinism.[19]

As a starting point, Critical Geopolitics highlights how geography is not solely a given objective, physical reality; it is also the product of a range of non-physical influences and perceptions, an on-going process, the means by which the state's policy makers make sense of its territorial setting, a constructed understanding—driven by two inter-related processes.[20] One is the process of 'visualization', of 'seeing' (that is, images), which is a culture-bound experience. This includes how one visualizes or experiences the physical space in which one is located (the terrain). The other is that of 'narration', of myth-making. How one perceives the physical space is also a product of one's strategic culture and military history, to cite two examples of 'narratives'. Geopolitics can thus be understood as the spatial practices, both material and representational, of statecraft itself, and hence the critical study of geopolitics focuses on the particular cultural mythologies or representations of the state.[21]

The geopolitics that informs national policies is thus the interplay between the 'practical geopolitics' of statecraft on the one hand, and 'formal geopolitics' on the other.[22] Classical Geopolitics as a particular mode of representing global space involves visualizing the world as a single picture, the transformation of time into space, state-centric representations of global space, and the pursuit of primacy by dominant states.[23] In other words, we 'see' the world around us, and that vision is the product of cultural and other influences. To return to the point suggested earlier about the geopolitics of conflicting territorial claims, what critical geopolitics does is to highlight that this is fundamentally about two conflicting geopolitical visions or cartographies. However, as noted earlier, the state appears as the end product of a naturalizing discourse.

Further, Critical Geopolitics argues that the world is 'spatialized' into regions with assigned attributes and characteristics. One such example of 'spatialization' is the idea of Southeast Asia. Despite the 'reality' of very different states, societies and cultures, the states of the Association of Southeast Asian Nations (ASEAN) have come to accept that the region comprised ten states—a so-called ASEAN 10—and hence, the objective from the start of ASEAN for a whole and complete ASEAN, comprising all the states of the region. When Cambodia became a fully-fledged member of the Association of Southeast Asian Nations (ASEAN) in 1999, the then-ASEAN Secretary-General Rodolfo Severino greeted this as "truly a historic moment," that this fulfilled the vision of ASEAN's "to unite all nations of Southeast Asia under one roof."[24]

As much as Critical Geopolitics adopted a dismissive attitude towards Classical Geopolitics, a recent third strand has sought to merge the two schools. Indeed, as early as 2005, as an Introduction to a special issue of the journal *Geopolitics*, Nicol and Minghi argued for a Geopolitics that could accommodate the emphasis on discursive practices and deconstruction in Critical Geopolitics on the one hand, with the traditional emphasis of Classical Geopolitics on the analysis of actual borders on the other.[25] In other words, there is a physical reality that may be culturally understood, it nevertheless exists as a physical reality whose existence cannot be "constructed" away.[26] South Korea cannot escape the physical reality that Seoul, its national capital, lies only thirty-odd kilometres from the De-Militarised Zone (DMZ). If there was a peace treaty between North and South Korea, and the DMZ regarded as a legal border between the two states, would this change South Korean perceptions of vulnerability vis a vis their northern neighbor? However, if there were an internationally recognized border in place of the DMZ, and a peace treaty between the two Korean states, this concern would likely recede. The significance of the distance between the DMZ and Seoul is, in other words, a human construct as much as it is a physical, unchanging reality. It may seem, from the Korean example, that it is ultimately the constructivist perspective that matters, that the thirty-kilometre distance between Seoul and the

DMZ matters precisely because the Republic of Korea—hereafter, ROK—has come to identify the Democratic Republic of Korea—hereafter, DPRK- as the putative adversary, and the DMZ becomes the front-line. However, the point that this chapter makes here is that short of ROK elites being able to 'construct' away the putative threat from the DPRK, the thirty-kilometre distance and the corresponding lack of strategic depth has become ontologically real for Seoul, and materially substantiated in a way that similar claims from Pyongyang would not be.

What do Classical and Critical Geopolitics bring to defense policy and strategy? This will be discussed later in this chapter. For the moment, suffice it to say that the relationships between states are, to a large extent, about matters of geography, both in its material and representational senses. Inasmuch as states argue over pieces of physical geography, it forms the context of particular inter-state relations. These pieces of real estate form the intellectual image that policy-makers have of their state. Territorial disputes are therefore as much contests over actual pieces of real estate as they are contests of conflicting mental images. Where the images and mental cartographies of the state do not coincide with the actual physical extent of the state, the potential for inter-state conflict therefore exists. Geopolitics therefore sets the context and condition for inter-state rivalry and conflict. Mutually hostile states thereafter form strategies that address these particular geopolitical questions. The next section examines the influence of geography on military operations, and will emphasize how geography provides both opportunities and constraints on the extent of military operations. How policy-makers weigh these opportunities and constraints then influences the stability of inter-state relations, and influences the stability of the strategies that these states employ when dealing with these questions about geography.

Geostrategy

At first glance, the relationship between geography and strategy ought to seem obvious. If the subject of geography is the study of "what the environment is like at any given place and time," then military geography, or Geostrategy, "concentrates on the influence of physical and cultural environments over political-military policies, plans, programs, and combat/support operations of all types in global, regional, and local contexts."[27] As with Classical and Critical Geopolitics, Geostrategy incorporates both the physical and cultural environments, which means Geostrategy appears to accept the proposition that the human construction of meaning in understanding of the physical realm is important.

Geography can affect strategy at three levels. in three ways. Firstly, it inspires the "grand narrative of high theory that appears as the common understanding of geopolitics."[28] At a geopolitical level, geography influences, amongst others, the identity of the belligerents, whose status as belligerents stems from the relative location and scope of their political 'space'. This is an issue that has already been explored. Furthermore, it provides the physical location and context of strategy. Geostrategy "places the planning and management of war in the context of geographical, physical and artificial (man-made) characteristics of the operational region."[29] The location of states is an important influence on the way policy makers and strategic planners think about strategy. Britain's pre-World War II strategy for protecting its imperial interests in Asia and the Pacific hinged on the location of a suitable naval base to which the Royal Navy, homeported in the United Kingdom, could deploy a fleet when necessary. Given that British naval planners had identified Japan as the putative adversary, this ruled out three options: Hong Kong, for being too close, and Sydney and Trincomalee for being too far away; this left the fourth option, Singapore, as the only one neither too far not too near to the putative adversary. Historically, Great Britain and the United States, being physical separated from the European landmass, could be less unconcerned and more detached about continental affairs. As a result, British and US strategic doctrines tilted towards air or naval theories at the expense of theories of land warfare.[30] Germany, and its historical antecedent, Prussia, could not escape the reality that it was sandwiched between larger, and potentially more powerful Great Powers in the form of France and Russia.[31]

At the second level, geography influences—it may even drive—strategic planning in terms of specific military operations, the types of forces required, as well as the logistics needed to support such operations. The physical extent of the geography of the borders shapes how defense planners regard the defensibility of those borders, and the subsequent adoption of either offensive or defensive strategic postures.[32] Physical size is typically regarded as determining the extent of a state's strategic depth; rugged and uneven geography tends to be seen as defensible, whereas flat open geography presents more options for offensive operations; these are the geographical factors that necessarily exist as part of strategic planners' calculations.[33] Rigorous war-gaming and study of the Pacific Ocean in the context of a hypothetical war against Japan bore fruit eventually in the Central Pacific counter-offensive that resulted in the re-capture of the Marianas islands from which B-29 bombing attacks on Japan would culminate in the nuclear bombing of Hiroshima and Nagasaki. For much of the Cold War, North Atlantic Treaty Organisation (NATO) planners obsessed over how they could stop a hypothetical Warsaw Pact conventional offensive at the Fulda Gap, universally regarded as the only defensive geography of the European theatre of operations. Israel's small size and resultant lack of strategic depth arguably necessitated pre-emptive conventional military operations against any military operations launched by its Arab neighbors.

Finally, geography will exercise a direct impact on the actual conduct and outcome of military operations; at least that is the prevailing view.[34] One scholar puts it bluntly: "Ignoring geography or the incorrect analysis of geography has often led to military disasters."[35] As suggested above, if it is axiomatic that defensive operations enjoy a natural advantage over offensive operations, a major aspect of military planning will be focused on the identification of types of geography that actually favor such defensive operations. One such geographical type is high ground, combined with the absence of cover and concealment. Another geographical type that influences the conduct of military operations is the presence of multiple potential lines of military operations. The presence of multiple potential lines of operations complicates defensive operations; however, if these multiple lines coalesce around a hub, such as a town that serves as a junction for the road and railway networks, this location will almost certainly become the focus of military operations for both the defensive and offensive. How the geography favors either offensive or defensive operations will have a bearing on calculations about the efficacy and desirability of military operations. Inasmuch as policy makers conclude that the geography does not provide the optimum environment for the use of military force to attain particular state goals, there is a good chance the military instrument will not be utilized.

It ought to be obvious that the above discussions of geostrategy veer dangerously close to the potential fallacy of geographical determinism; simply put, there are no concrete rules about the offensive or defensive merits of particular geographical types.[36] Almost always, assessments about the geostrategic ramifications of specific types of geography tend to unravel upon closer examination. Deserts are often regarded as ideally suited for armored operations, but the reality is that fine, loose sand can actually impede the movement of even tracked vehicles.[37] Similarly, mountainous geography may appear to favor the defense over offensive operations: after all, the geography affords the defender interlocking fields of fire and natural obstacles to the mobility of the attacking force; nevertheless, there are disadvantages to the defense such that defense does not necessarily translate into tactical, operational or strategic advantages.[38] The geography of the jungle may favor small unit action and prohibit the use of mobile forces, limit visibility and fire support, and make C[3]I (command, control, communications, intelligence) highly problematic; but the famous axiom, that the jungle is neutral, means that endurance and stamina—the key requirements for victory—are in short supply for both defender and attacker. The point of this discussion is that strategic planners have to remember that the geographic influences on military operations remain fluid and open to interpretation.

If cultural mindsets and human perception are inevitable influences on Geopolitics, they are similarly inevitable in Geostrategy. Policy makers and strategic planners perceive the tactical, operational and strategic ramifications of the geographical features of the arena of military operations. British strategic planning for the pre-World War II defense of the Singapore Naval Base—Operation Matador[39]—hinged on what they understood to be the only conclusion that could be drawn from the geography of the Malay peninsula to the north of Singapore: that even if Japan were able to land forces anywhere along the Malay peninsula, the tropical jungles—which these planners believed to be the singular feature of the geography of the peninsula—would act as a natural operational and tactical barrier that could only serve to assist British defenders of the Singapore Naval Base. The subsequent experience of the Japanese campaign showed these beliefs to be false. However, these false beliefs were politically convenient and expedient for British policy makers in London: the mistaken belief that the jungles of the Malay peninsula provided an illusion of a workable solution to the problem of insufficient resources for the protection of the entire British empire.

Geopolitics and Geostrategy—The Basis of Defense Policy and Strategy

If statehood is, as earlier argued, a claim to exclusive ownership of a given piece of real estate, then it is intuitive to expect that geography is of fundamental importance to strategy and defense policy. Singapore's existence as a sovereign state is fundamentally connected to its physical and human geography: the small size of the island and its corresponding lack of natural resources; its location astride one of the global economy's most important sea lines of communication; the geographical fact that, despite being an archipelago of small islands, it is almost completely surrounded by the territorial waters of Malaysia and Indonesia to its north and south respectively. Surely Singapore's policy makers and strategic planners cannot escape these geographical "hard truths" in the crafting of the state's defense policies and strategies.[40]

In other words, it is important not to be blind to the influence that the physical environment can have on states and their policy makers and strategic planners. The objectives that a state's policy makers focus on, the decisions they make to secure these objectives, exist in a geographical context, and often are geographical in nature. In so doing, these decisions can shape the relationships that the state has with its neighbors. Singapore's relationship with Malaysia was shaped by the simple geographical fact that until fairly recently, Singapore depended on Malaysia for much of its potable water. The decisions that Singapore's policy makers took to secure the supply of potable water from Malaysia—including enunciating a strategic doctrine that envisaged the use of military force where necessary—therefore fundamentally shaped the bilateral relationship.[41] Geopolitics, in particular Classical Geopolitics, therefore helps policy makers and strategic planners to identify the points of potential contention in the state's relationships with other states, and especially contiguous neighbors. In so doing, what can emerge is a theory of war for the particular state: an understanding of how and why the state may go to war with another state.[42]

At the same time, however, a less gloomy picture can emerge from a careful study of the geography of the state. A theory of geopolitical interconnectedness suggests that the potential for conflict between states can be mitigated by the existence of inter-connectedness between the particular states and their societies.[43] Relationships between contiguous states do not have to be conflictual in nature, and for every relationship that contains latent conflict, there are any number of other contiguous relationships that are essentially peaceful. Even conflictual relationships can transform, as witnessed by the signing of peace treaties between the modern state of Israel and a number of its previously hostile Arab neighbors.

This discussion should have pointed to the importance of non-objective considerations that something as apparently objective as geographical features might introduce into defense policy and strategy. It may be true that inter-state relationships are shaped by the policies and strategies of the respective states in pursuit of their own national goals, and the extent of contradiction and frustration over the attainment of these goals; it may even be true that territorial disputes are the root cause of many conflicts.[44] However, it is important to recognize pre-decision mental processes, especially with regards to perceptions of environment and space leading to mental maps, intra and inter-group perceptions and images.[45] Herein lies the value of Critical Geopolitics, because it reminds us that these conflicts are 'real' only because human actors—policy makers and their strategic planners working in support—choose to identify specific geographies as fundamental to their state's national interest. Borders, for instance, are intellectual constructs: "National states exist almost entirely in the minds of the human race ... Their frontiers are for the most part invisible from outer space."[46] If Classical Geopolitics claims that "its dominant modes of narration are declarative (this is how the world is) and imperative (this is what we must do)"[47], Critical Geopolitics demonstrates that the first assumption can be regarded as problematic at least, if not outright false.

That is not to deny the influence that geography can exercise on defense policy and strategy. Soldiers in the midst of a military operation likely have neither the time nor the inclination to realize that the geographical features that they are fighting to capture have no intrinsic political or strategic value, that these 'values' are as much 'imagined' as they are 'real'. The strategists who plan these operations and the policy makers who approve them, similarly, will not agonize over the extent to which the geostrategic significance of a particular geographical feature is imagined or otherwise. Furthermore, military history will often demonstrate how the ability to assert control over certain territorial positions will have a major impact on the military effort. If the Singapore Armed Forces intends to launch a pre-emptive military operation against its putative adversary, because Singapore's policy makers and strategic planners conclude that the island lacks geostrategic depth, its military planners will not worry about the extent to which this 'lack of geostrategic depth' is imagined.[48] The evolution of the Singapore Armed Forces is thus a case study of the impact of the geopolitical and geostrategic conditions of Singapore—its small geographical size, its prior dependence on external sources for its potable water, its demographic patterns especially in relation to its northern and southern contiguous neighbors—on the organization's strategic thinking, operational doctrines and force structures.[49]

Conclusions

What this chapter has tried to show is that geography exercises a two-level influence on the strategic calculus of the state. One, it provides the policy makers and strategic planners the context of the state and its relationships—especially antagonistic, if not outright hostile relationships—with other states both contiguous and further afield. More often than not, mutually hostile states are so precisely because of some geopolitical issue between them, usually issues of territorial control, but also of resource control. States in so-called conflict zones are mutually hostile not because of some magical unseen force exerted by the geography of the region. Rather, the potential for conflict resides in how the real estate of the region has been divided between those states, or in the relative imbalance of natural resource allocations, or in the way that the particular geography has encouraged or hindered movement from one place to the next.

Two, it informs those policy makers and strategic planners of the likely scenarios in which the state will be embroiled in armed hostilities, the extent to which the 'objective geography' imposes intrinsic vulnerabilities, and the subsequent policies and strategies that these policy makers and strategic planners will have to put in place to address these vulnerabilities. Geography will influence strategic calculations about relative geostrategic opportunities and constraints and by making the recourse to the military instrument seem either viable or even urgent; in so doing, it influences policy-makers' decisions about the employment of the military instrument in the pursuit of state objectives. The earlier discussion demonstrates how specific physical geography either facilitates or hinders specific types of military operations. Both policy makers and strategic planners must factor these into their calculations about the feasibility of using military instruments in the pursuit of national political objectives against another state.[50]

However, these assessments do not necessarily exercise a deterministic influence on the policies and strategies adopted thereafter. Whether real or otherwise, geostrategic vulnerabilities do not guarantee that the state's strategic planners and policy makers will necessarily gravitate towards pre-emptive strategies. In the early 1970s, South Korea's strategic planners and policy makers faced a potentially bleak security outlook: the United States enunciated the so-called Nixon Doctrine, which suggested that South Korea's principal security guarantor was reducing its security commitment; however, North Korea was, then, still militarily more powerful, and had engaged in a series of provocative actions, including an attempted assassination of South Korean President Park Chung Hee in January 1968; finally, South Korea's attempts to strengthen its military instrument were being stymied by the United States, because the latter did not want to further destabilize the strategic situation on the Korean peninsula. The conundrum was how South Korea ought to proceed: build up its military power by finding alternative sources of military hardware, and in the process, risk further exacerbating tensions with North Korea; or, seek to de-escalate tension, either by choosing to not respond to North Korean provocations at the very least, or through rapprochement with its northern neighbor. South Korea opted for de-escalation.

In other words, policy makers have options available; and, even if strategic planners recommend military actions, policy makers do not have to accept these recommendations. What geography does for the defense policies and strategies of any state is that it sets the stage and the context, it identifies the interests that the state's policy makers and strategic planners will focus on, and it identifies the opportunities and limitations within which those defense policies and strategies have to operate in.

However, to use the analogy of a pair of spectacles with which a person may use to see clearly, geography provides merely one lens in this pair of spectacles. Just as binocular vision provides depth perception, the pair of spectacles requires a second lens which, when combined with the geography lens, allows the state's policy makers and strategic planners a more detailed and nuanced view of the environment in which the state exists. This other lens that is needed is provided by the strategic culture of the state. This will be the focus of the next chapter.

Endnotes

[1] Emily O. Goldman, "Thinking About Strategy Absent the Enemy," *Security Studies*, vol. 4, no. 1, 1994, pp. 40–85.
[2] On geopolitics, see, for instance: John Agnew, *Geopolitics: Re-visioning World Politics* (London and New York: Routledge, 1998); Priya Chacko (ed.), *New*

Regional Geopolitics in the Indo-Pacific: Drivers, Dynamics and Consequences (London and New York: Routledge, 2016); Klaus Dodds and Mark Nuttall, *The Scramble for the Poles: The Geopolitics of the Arctic and Antarctic* (Cambridge and Malden: Polity Press, 2016); Gearóid Ó Tuathail, *Critical Geopolitics: The Politics of Writing Global Space* (London: Routledge, 1996); Geoffrey Sloan, *Geopolitics, Geography and Strategic History* (London and New York: Routledge, 2017). On geostrategy, see, for instance: John M Collins, *Military Geography for Professionals and the Public* (Washington, DC: Brassey's, 1998); Hugh Faringdon, *Confrontation: The Strategic Geography of NATO and the Warsaw Pact* (London: Routledge & Kegan Paul, 1986); Colin S. Gray and Geoffrey Sloan (eds.) *Geopolitics, Geography and Strategy* (London and Portland: Frank Cass, 1999); Louis C. Peltier and G. Etzel Pearcy, *Military Geography* (Princeton: Van Nostrand, 1966).

[3] Colin S. Gray, "Inescapable Geography," in Colin S. Gray and Geoffrey Sloan (eds.), *Geopolitics, Geography and Strategy* (London and Portland: Frank Cass, 1999), p. 164.

[4] For Critical Geopolitics, see, for instance: Mat Coleman, "Geopolitics as a Social Movement: The Causal Primacy of Ideas," *Geopolitics*, vol. 9, no. 2, Summer 2004, pp. 484–491; Carl Dahlman and Stanley Brunn, "Reading geo-politics beyond the state: organisational discourse in response to 11 September," *Geopolitics*, vol. 8, no. 3, Autumn 2003, pp. 253–280; Simon Dalby, "Imperialism, Domination, Culture: The Continued Relevance of Critical Geopolitics," *Geopolitics*, vol. 13, no. 3, Autumn 2008, pp. 413–436; Sami Moisio and Anssi Paasi, "Beyond State-Centricity: Geopolitics of Changing State Spaces," *Geopolitics*, vol. 18, no. 2, Summer 2013, pp. 255–266; Oliver Turner, "China, India and the US Rebalance to the Asia Pacific: The Geopolitics of Rising Identities," *Geopolitics*, vol. 21, no. 4, Winter 2016, pp. 922–944; Albert Tzeng, William L. Richter and Ekaterina Koldunova (eds.), *Framing Asian Studies: Geopolitics and Institutions* (Singapore: ISEAS Publishing, 2018). For Classical/ Neo-Classical Geopolitics, see, for instance: Terrence Haverluk, Terrence W, Kevin M. Beauchemin and Brandon A. Mueller, "The Three Critical Flaws of Critical Geopolitics: Towards a Neo-Classical Geopolitics," *Geopolitics*, vol. 19, no. 1, Spring 2014, pp. 19–39; Phil Kelly, *Classical Geopolitics: A New Analytical Model* (Stanford: Stanford University Press, 2016); Heather M. Nicol and Julian Minghi. "The Continuing Relevance of Borders in Contemporary Contexts," *Geopolitics*, vol. 10, no. 4, Winter 2005, pp. 680–687; Benjamin Schreer, "Towards Contested 'Spheres of Influence' in the Western Pacific: Rising China, Classical Geopolitics, and Asia-Pacific Stability," *Geopolitics*, vol. 24, no. 2, Summer 2019, pp. 503–522.

[5] Colin S. Gray, *Modern Strategy* (Oxford: Oxford University Press, 1999), pp. 1–2.

[6] Ewan Anderson, "Geopolitics: International Boundaries as Fighting Places," in Colin S. Gray and Geoffrey Sloan (eds.), *Geopolitics, Geography and Strategy*

(London and Portland: Frank Cass, 1999), p. 125. Also see: Andrew Kirby, "The Construction of Geopolitical Images: the world according to Biggles (and other fictional characters)," in Klaus Dodds and David Atkinson (eds.), *Geopolitical Traditions: critical histories of a century of political thought* (London: Routledge, 2000), p. 52.

[7] Geoffrey Sloan, "Sir Halford J. Mackinder: The Heartland Theory Then and Now," in Colin S. Gray and Geoffrey Sloan (eds.), *Geopolitics, Geography and Strategy* (London and Portland: Frank Cass, 1999), 15–17.

[8] Roy E.H. Mellor, *National Defence: The Military Aspects of Political Geography—A reconnaissance study* (Aberdeen: Department of Geography, University of Aberdeen, 1987), pp. 22–24.

[9] Louis C. Peltier and G. Etzel Pearcy, *Military Geography* (Princeton: Van Nostrand, 1966), p. 40.

[10] John O'Loughlin and Luc Anselin, "Geography of International Conflict and Cooperation," in Michael Don Ward (ed.), *The New Geopolitics* (Philadelphia: Gordon & Breach, 1992), p. 12.

[11] Kook-chin Kim, "The Pivotal Security Linkage Between the Korean Peninsula and Northeast Asia," *Korea and World Affairs*, vol. 9, no. 3, 1985, pp. 491–492.

[12] Bernard Fook Weng Loo, "Goh Keng Swee and the Emergence of a Modern SAF: The Rearing of a Poisonous Shrimp," in Emrys Chew and Kwa Chong Guan (eds.), *Goh Keng Swee: A Legacy of Public Service* (Singapore: World Scientific, 2012), pp. 132–133.

[13] Gray, *Modern Strategy*, p. 3.

[14] Mellor, *National Defence*, p. 25.

[15] See, for instance: Simon Dalby, "Calling 911: geopolitics, security and America's new war," *Geopolitics*, vol. 8, no. 3, Autumn 2003, pp. 61–86.

[16] Schreer, "Towards Contested 'Spheres of Influence' in the Western Pacific."

[17] For instance, see: Tuomas Forsberg, "The Ground without Foundation: Territory as a Social Construct," *Geopolitics*, vol. 8, no. 2, Summer 2003, pp. 7–24; Gearóid Ó Tuathail, *Critical Geopolitics: The Politics of Writing Global Space* (London: Routledge, 1996).

[18] J.N.H. Douglas, "Conflict Between States," in Michael Pacione (ed.), *Progress in Political Geography* (London: Croom Helm, 1985), p. 88; Joe Painter, *Politics, Geography and Political Geography: A Critical Perspective* (London: E. Arnold; New York: Halsted Press, 1995) pp. 28–55.

[19] Jan Nijman, "The Dynamics of Superpower Spheres of Influence: U.S. and Soviet Military Activities, 1948–1978," in Michael Don Ward (ed.), *The New Geopolitics* (Philadelphia: Gordon & Breach, 1992), pp. 91–119.

[20] Ó Tuathail, *Critical Geopolitics*, pp. 68–140.

[21] Nurit Kliot, "The Political Geography of Conflict and Peace—An Introduction," in Nurit Kliot and Stanley Waterman (eds.), *The Political Geography of Conflict and Peace* (London: Belhaven Press, 1991), pp. 8–9; Gearóid Ó Tuathail and

Simon Dalby, "Rethinking Geopolitics: Towards a critical geopolitics," in Gearóid Ó Tuathail, and Simon Dalby (eds.), *Rethinking Geopolitics* (New York: Routledge, 1998), p. 3.

[22] Gearóid Ó Tuathail and James Agnew. "Geopolitics and discourse: Practical geopolitical reasoning in American foreign policy," *Political Geography Quarterly*, 11, 1992, p. 10.

[23] Gearóid Ó Tuathail, "Postmodern Geopolitics? The modern geopolitical imagination and beyond," in Gearóid Ó Tuathail, and Simon Dalby (eds.), *Rethinking Geopolitics* (New York: Routledge, 1998), pp. 20–23.

[24] https://www.khmertimeskh.com/599143/what-it-means-for-cambodia-to-be-an-asean-member-state/.

[25] Heather M. Nicol and Julian Minghi. "The Continuing Relevance of Borders in Contemporary Contexts," *Geopolitics*, vol. 10, no. 4, Winter 2005, pp. 680–687. Also see: Colin Flint and Virginie Mamadouh, "The Multi-Disciplinary Reclamation of Geopolitics: New Opportunities and Challenges," *Geopolitics*, vol. 20, no. 1, pp. 1–3; Terrence Haverluk, Terrence W, Kevin M. Beauchemin and Brandon A. Mueller, "The Three Critical Flaws of Critical Geopolitics: Towards a Neo-Classical Geopolitics," *Geopolitics*, vol. 19, no. 1, Spring 2014, pp. 19–39.

[26] Phil Kelly, "A Critique of Critical Geopolitics," *Geopolitics*, vol. 11, no. 1, Spring 2006, pp. 24–53. Also see: Bernard F.W. Loo, *Middle Powers and Accidental Wars: A Study in Conventional Strategic Stability* (Lewiston, NY: Edwin Mellen Press, 2005), pp. 69–104.

[27] Collins, *Military Geography for Professionals and the Public*, p. 3.

[28] Gray, "Inescapable Geography," p. 165.

[29] Tal Tovy, *The Changing Nature of Geostrategy 1900–2000* (Maxwell AFB, Alabama: Air University Press, 2015), p. xv.

[30] Williamson Murray, "Some Thoughts on War and Geography," in Colin S. Gray and Geoffrey Sloan (eds.), *Geopolitics, Geography and Strategy* (London and Portland: Frank Cass, 1999), pp. 206–211.

[31] Hugh Faringdon, *Confrontation: The Strategic Geography of NATO and the Warsaw Pact* (London: Routledge & Kegan Paul, 1986), pp. 74–75.

[32] Murray, "Some Thoughts on War and Geography," pp. 211–215.

[33] Collins, *Military Geography for Professionals and the Public*, pp. 14–21; Gray, "Inescapable Geography," pp. 171–172; Mellor, *National Defence*, pp. 20–21, 40–79; Patrick O'Sullivan, *Terrain and Tactics* (New York, Westport, and London: Greenwood Press, 1991), pp. 9–30; Patrick O'Sullivan and Jesse W. Miller, *The Geography of Warfare* (London: Croom Helm, 1981), p. 81.

[34] Trevor N. Dupuy, *Understanding War: History and Theory of Combat* (London: Leo Cooper, 1987), pp. 160–163. Also see: Collins, *Military Geography for Professionals and the Public*; O'Sullivan and Miller, *The Geography of Warfare*.

[35] Tovy, *The Changing Nature of Geostrategy 1900–2000*, p. 27.

[36] Collins, John M. Collins, Military Geography for Professionals and the Public (Washington, D.C.: Brassey's.1998). Louis C. Peltier and G. Etzel Pearcy, Military Geography (Princeton: Van Nostrand, 1966).

[37] Patrick O'Sullivan, "The Geopolitics of Deterrence," in David Pepper and Alan Jenkins (eds.), *The Geography of Peace and War* (Oxford: Blackwell, 1985), p. 67.

[38] Such close terrain only staves off defeat for the defender and buys time. At some point, the defender has to emerge from this close terrain and assume a counter-offensive against its adversary. The Chinese Communist Party could degrade the relative combat power of its Kuomintang adversary while the former was ensconced in the mountains, but it was only the third phase, the strategic counter-offensive, that ultimately gave the Chinese Communists strategic success.

[39] Chit Chung Ong, *Operation Matador: Britain's war plans against the Japanese, 1918–1941* (Singapore: Times Academic Press, 1997).

[40] Lee Kuan Yew, *Hard Truths to Keep Singapore Going* (Singapore: Straits Times Press, 2011). The point to note is that these so-called "hard truths" are presented and understood in a manner that proponents of Classical Geopolitics will easily identify with: as inescapable realities that Singapore policy makers have to navigate and negotiate, and the Singapore electorate cannot but accept the sacrifices that these policy makers' choices will impose on the electorate.

[41] Norman Vasu and Bernard F.W. Loo, "National Security and Singapore: An Assessment," in Terence Chong (ed.), *Management of Success* (Singapore: Institute of Southeast Asian Studies, 2010), pp. 462–486.

[42] Paul F. Diehl, "Geography and War: A Review and Assessment of the Empirical Literature," in Michael Don Ward (ed.), *The New Geopolitics* (Philadelphia: Gordon & Breach, 1992); John O'Loughlin and Herman van der Wusten, "Political geography of war and peace, in Peter J. Taylor (ed.), *Political Geography of the Twentieth Century: a global analysis* (London: Belhaven Press, 1993); Harvey Starr, "Joining Political and Geographical Perspectives: Geopolitics and International Relations," in Michael Don Ward (ed.), *The New Geopolitics*. Philadelphia: Gordon & Breach, 1992).

[43] O'Loughlin and van der Wusten, "Political geography of war and peace," pp. 84–86. The inter-connectedness argument can be used to describe the Malaysia-Singapore relationship, where the two states and societies are inter-connected at political, economic, societal and familial levels; see, for instance: Saw Swee Hock and K. Kesavapany (eds.), *Singapore-Malaysia Relations Under Abdullah Badawi* (Singapore: Institute of Southeast Asian Studies Press, 2006).

[44] Douglas, "Conflict Between States," p. 83; O'Loughlin and van der Wusten, "Political geography of war and peace," p. 104.

[45] Douglas, "Conflict Between States," pp. 99–100.

[46] Kenneth E. Boulding, "The Nature and Cause of National and Military Self-Images in Relation to War and Peace," in Nurit Kliot and Stanley Waterman (eds.), *The Political Geography of Conflict and Peace* (London: Belhaven Press, 1991), p. 142.

[47] Sloan and Gray, "Why Geopolitics?," p. 5.

[48] Tim Huxley has argued that a pre-emptive military operation against the adversary is the only option available to Singapore; see: Tim Huxley, *Defending the Lion City: The Armed Forces of Singapore* (Crows Nest, NSW: Allen & Unwin, 2000). Also see: Richard A. Deck, "Singapore: Comprehensive Security—Total Defence," in *Strategic Cultures in the Asia-Pacific Region* Ken Booth and Russell Trood (eds.) (Houndmills: Macmillan Press, 1999), pp. 248–49.

[49] Bernard F.W. Loo, "From Poisoned Shrimp to Porcupine to Dolphin: Cultural and Geographic Perspectives of the Evolution of Singapore's Strategic Posture," in Amitav Acharya and Lee Lai To (eds.), *Asia in the New Millennium: APISA First Congress Proceedings, 27–30 November 2003* (Singapore: Marshall Cavendish Academic, 2004), pp. 352–375; Bernard F.W. Loo, "Maturing the Singapore Armed Forces: from Poisonous Shrimp to Dolphin," in Bridget Welsh, James Chin, Arun Mahizhnan and Tan Tarn How (eds.), *Impressions: The Goh Years in Singapore* (Singapore: National University of Singapore Press, 2009), pp. 178–197.

[50] Benjamin A. Most, Harvey Starr, and Randolph M. Siverson, "The Logic and Study of the Diffusion of International Conflict," in Manus I. Midlarsky (ed.), *Handbook of War Studies* (Ann Arbor: University of Michigan Press, 1993) p. 121.

Chapter 2

Strategic Culture and Strategy

Bernard F.W. Loo

Senior Fellow, Institute of Defence and Strategic Studies,
S Rajaratnam School of Internal Studies,
Nanyang Technological University, Singapore

isfwloo@ntu.edu.sg

This chapter forwards the argument that the strategic choices made by the policy makers of any state, manifested in the state's defense policies and strategies, derive from the milieu of space and time in which the state exists. The preceding chapter has addressed one aspect of the element of space in the milieu of the state, namely, the geography—especially the physical and human geography of the state and the region in which it exists—and how this geography relates to the defense policies and strategies of the state. When geography combines with time—in other words, the history of the state—these elements help to create the strategic culture of the state. To summarize, this chapter argues that the strategic culture of the state will exercise a strong influence on the strategic choices that the state's policy makers will adopt, in pursuit of the state's national interests.[1] As the strategist Colin Gray, and one of the chief proponents of strategic culture theory had argued, strategy (and defense policy, for that matter) is not solely driven by objective facts, and the strategic calculus is not an exclusively rational exercise. Rather, strategy and defense policy are made by human beings, and human beings necessarily exist in a pervasive wider cultural context that influences all forms of human action and interaction.[2]

It is necessary to begin with a caveat: it is difficult to study the effect that something ephemeral like strategic culture can have on strategic behavior. As Richard Ned Lebow argued, it can be difficult to reconstruct how policy makers perceive events and developments, and it is potentially problematic to suggest that there is a causal connection between those perceptions and the eventual strategic choices made.[3] These methodological problems make the strategic culture proposition appear as no more than after-the-fact explanations—and worse, justifications—of existing strategic behavior.[4] Charles Kupchan outlined three objections to the strategic culture proposition: one, culture tends to be used as a residual variable, that is, it is used when other more conventional and concrete variables fail to provide an adequate explanation of the outcome; two, culture is an amorphous and malleable concept that is difficult to define and to measure; and, three, inasmuch as culture changes little over time, critics claim that cultural arguments tend to be non-falsifiable.[5] Finally, echoing the anthropological definition of culture forwarded by Edward Tyler, a criticism of the strategic culture argument posits that, over time, the international system will show what strategic behavior works and what does not; it then makes sense for policy makers, irrespective of their specific location in space and time, to gravitate towards those strategic behaviors that have been proven to work, and in the process create a common set of strategic practices, and consequently render cultural differences irrelevant.[6]

In other words, it can be dangerous to attribute too much explanatory power of strategic behavior to strategic culture. By itself, strategic culture is not a sufficient explanation of strategic behavior.[7] This need for caution notwithstanding, this chapter suggests that there is value in the strategic culture argument, for two reasons. One, by focusing on images and symbols, strategic culture "provides a deeply embedded notion of what constitutes national security, not a short-term set of suppositions about the nature of international relations."[8] Strategic culture can be a concept that is analytically powerful. Specific conceptions of national security can be tested to see if they take root within a political entity over an extended period and reshape the boundaries of politically legitimate behavior as well as the missions of the bureaucracy and military. These images may be able to explain variations in state behavior over the period in question.[9] Two, it directs our attention to how policy makers think about the identity of their state, and its commensurate roles and responsibilities; how these prescribe what these policy makers see as politically legitimate behavior; and how these intellectual paradigms and values subsequently influence these policy makers' strategic choices. about various options, and the strategic objectives around which national institutions such as the military define their roles and missions.

This chapter will thus begin with a discussion of the concept of strategic culture and its component elements, which it proposes comprises of three elements. One is the physical context on the state, that is, the physical and political geography of the state and the region in which it exists; this has already been addressed in the previous chapter. Hence this chapter will focus on the other two elements that combine to constitute the strategic culture of the state: the human geography and anthropological contexts of the state; and the history of the state and its peoples. Finally, it will address how strategic culture exercises its influence over the defense policies and strategies of the state.

Conceptualising Strategic Culture

The concept of strategic culture was first proposed by Jack Snyder in a RAND report published in 1977. Focusing on the Soviet Union, Snyder argued that the Soviet Union did not approach strategy, and in particular, nuclear strategy, in the same manner as did the United States. At that time, the United States strategic nuclear planners had adopted so-called "flexible response" doctrines that posited a controlled and limited nuclear war with the Soviet Union. Snyder argued, instead, that the Soviet Union's "unique historical experiences, distinctive political and institutional relationships, and a preoccupation with strategic dilemmas ... have combined to produce a unique mix of strategic beliefs and a unique pattern of behaviour based on those beliefs."[10] Simply put, how the two superpowers conceptualized nuclear weapons and their respective strategies for the use (or non-use, for that matter) were fundamentally different, and these differences stemmed from their different histories, political structures and strategic circumstances.

It is important to remember that Snyder conceived of strategic culture as a concept to explain peculiarities in a particular state's strategic behavior that other existing theories and concepts could not begin to explain.[11] This is because of the manner in which Snyder's original concept came to be understood and developed further by other strategists. The late British-American strategic thinker, Colin S. Gray, for instance, had taken

Snyder's argument about the uniqueness of the Soviet Union's understanding of the nature and strategic utility of nuclear weapons, applied this concept, albeit with a few modifications, to the United States, and made three claims[12]:

- There is a uniquely "American (and, *ab extensio*, other) strategic culture—which flows from geopolitical, historical, economic and other unique influences."
- It "provides the milieu within which strategic ideas and defense policy decisions are debated and decided.
- Finally, "an understanding of American strategic culture ... can help explain why American policymakers have made the decisions they have [and even] predict decisions in the future."

It is worth noting that Gray would later step back from that last aspect of his argument—the idea of being able to predict future decision—but Gray would continue to insist that strategic culture could help us understand existing strategic behavior.[13]

Another scholar who developed a strategic culture argument was Ken Booth. Two years after Snyder's RAND report was published, Ken Booth published *Strategy and Ethnocentrism*, which highlighted the influence of culture on one's own strategic perceptions and subsequent decisions. Booth understood ethnocentrism to have three meanings: it is a concept to describe perceptions of "group centrality and superiority"; it is a "faulty methodology" where scholars and policy makers project their own frames of reference onto others, and "assess aspects of other cultures in terms of one's own culture"; and, as a synonym of being 'culture-bound', in that they are unable to "see the world through the eyes of a different national or ethnic group."[14] Booth subsequently stated, "To 'know the enemy' has always been a cardinal tenet of strategy. If this goal is to be achieved in the future with more regularity than it has been achieved in the past, then cultural relativism should take its place in the strategist's lexicon."[15] In a later publication, Booth would define strategic culture as "a nation's traditions, values, attitudes, patterns of behaviour, habits, customs, achievement and particular ways of adapting to the environment and solving problems with respect to the use of force."[16]

A later scholar, Alistair Iain Johnston, constructed a framework that identified three generations of strategic culture theorists, and reserved his harshest assessments for the first generation theorists, including Jack Snyder and Colin Gray.[17] First-generation strategic culture theory's 'everything

including the kitchen sink' definition was simultaneously overly deterministic and too broad and made the concept "practically meaningless," and
over-simplified "often complex domestic influences on foreign policy."
Echoing those he identified as second-generation theorists,[18] Johnston further argued that first-generation theory failed to "acknowledge the factor of
instrumentality—that leaders could consciously cite strategic culture to
justify their decisions." Johnston, who saw himself as part of the third generation of theorists, argued that strategic culture needed to be more narrowly
defined, by focusing specifically on preferences and omitting strategic
behavior; hence, his definition of strategic culture as "an integrated system
of **symbols** (i.e., argumentation structures, languages, analogies, metaphors,
etc.) that acts to establish pervasive and long-lasting **grand strategic preferences** by formulating concepts of the role and efficacy of military force in
interstate political affairs, and by clothing these conceptions with such an
aura of factuality that the **strategic choices seem uniquely realistic and
efficacious** (emphasis mine)."[19] In so doing, Johnston argued that strategic
culture theory could be falsifiable, and thereby become more academically
rigorous and useful.

The main problem with Johnston's approach is a single concept, 'strategic culture', is transformed into two: 'strategic-culture'. Furthermore,
Johnston's model is ironically difficult to actually test empirically. Indeed,
as Alan Bloomfield argues, Johnston's model requires us to imagine a
"rather implausible, artificial and unrealistic" scenario of

> decision-makers sitting around a table deliberating about how to respond
> to a strategic crisis.
>
> When considering the evidence before them they would assign rela-
> tive weights to the material variables (geography, relative power, tech-
> nology, etc.) relevant to the situation at hand. We could the imagine the
> decision-makers *consciously* (emphasis original) doing the same for
> the strategic culture variable: they would, in effect, be measuring the
> strength of their collective strategic cultural preferences. They would
> then assess whether the combined weight of some of the material vari-
> ables overrode the weight of their strategic-cultural preferences, or
> whether other variables lent support to their strategic-cultural predispo-
> sitions. They would then choose from their list of possible responses to
> the crisis with rational, mathematical certainty.[20]

Not that this critique bothered Colin Gray, who rebuffed these concerns by making four observations.[21] One, as noted in the beginning of

this chapter, if strategy and defense policy is made by human beings, and human beings are inescapably cultural animals, it then necessarily follows that strategy has a cultural element or component to it. Two, Gray argued that what he and other so-called first-generation theorists were attempting to do was to remind scholars and policy makers that the adversary (in this case, the former Soviet Union) was not a "black-boxed superpower" that was beyond comprehension. Three, Johnston's quest for a falsifiable theory, by separating strategic behavior from strategic culture, "commits the same error as the doctor who sees people as having entirely separable bodies and minds."[22] Finally, the value proposition of strategic culture lies not in a belief that it can predict future behavior, but rather in its ability to help us understand strategic behavior in the present.[23] Sondhaus goes so far as to argue that Gray's "assertion that 'all dimensions of strategy are cultural' has been maddening to his critics precisely because it is so difficult to refute."[24] Furthermore, other scholars have taken issue with Johnston's quest for falsifiable theory as methodologically problematic, insofar as strategic culture cannot act independently, and therefore the effect of strategic culture on strategic behavior cannot be rigorously studied.[25]

The Gray-Johnston debate came to frame much of subsequent work on strategic culture. Most strategic culture theorists since have fallen into either the Gray or Johnston camps. A number of theorists and strategists have attempted to move the debate further. Some scholars have attempted to merge behavior and culture, by arguing that culture forms the context in which behavior occurs, or by suggesting that culture forms the lens through which behavior is interpreted and understood.[26] Other scholars saw merit in combining the contextual with the interpretivist arguments into a model that moves strategic culture closer to its roots in sociology, anthropology and psychology. In other words, strategic culture ought to be thought of as a "'toolbox' which contains interpretive 'strategies of action' or 'culturally available ways of organizing collective behaviour' ... 'cognitive schemas' [and] processing devices or cognitive shortcuts that help to order the mass of data which confronts us ... packages of ideas and behavioural patterns inextricably linked together [which] affect our perceptions of the world, our attribution of causality, and the attendant behavioural responses which are considered appropriate."[27]

This chapter argues that strategic culture is a necessary element of defense policy and strategy, for three reasons. One, it provides a kind of conceptual lens with which policy makers and strategic planners see themselves—that is, the state—and the world in which the state exists. conceptual lens is an ideational framework that comprises two main elements—the geography and history of the state—in which the physical and material realities of the state's geography and history have no innate political and strategic meaning other than that which the human actors that comprise the state confer on them. Borrowing from Johnston's definition of strategic culture mentioned earlier, it is the "symbols ... languages, analogies, metaphors," derived from the geographical and historical context of the state, that inform policy makers and strategic planners of the identity of the state, its place in the wider regional and international context that consists of other political actors.

Two, this conceptual lens allows these actors to identify the specific security interests of state, and the potential threats and challenges to these interests that will emanate from the environment in which the state exists. Furthermore, these actors can subsequently prioritize these interests, and the threats and challenges thereto. Finally, again, borrowing from the Johnston definition discussed earlier, these symbols subsequently suggest the "grand strategic preferences [that] seem uniquely realistic and efficacious." Once the various threats and challenges to the interests of the state are identified and prioritized, it surely follows that policy makers will tend towards and prefer the strategic instruments in their 'toolbox' that specifically address these threats and challenges. It is the inversion of the old adage—if the only tool you have is a hammer, you will start treating your problems like a nail—wherein if all you see are nails around you, you will tend to reach for the hammer in your toolbox. The case of Singapore is illustrative: because of its limited geographical size and historical experiences with its contiguous neighbors, Singapore's policy makers have traditionally seen potential vulnerabilities in its geopolitical neighborhood; as a consequence, Singapore's policy makers have historically placed great emphasis on developing the capacity to defend the island state against adversaries, real or imagined.[28]

Know Yourself, Know the Enemy—Strategic Culture as the Foundation of Policy and Strategy

The preceding discussion of the concept of strategic culture places great emphasis on the roles of policy makers, and strategic planners whose work supports the former. Strategic culture arguably constitutes a foundational influence on the defense policy and strategy of the state, because it shapes the perceptions of policy makers and strategic planners in four aspects: the identity of the state; the nature and character of the strategic environment; the identities and intentions of the putative adversaries, and; the level and immediacy of the threats posed to the state's national security interests.

To begin with, culture influences how policy makers construct the self-identity of their state. How policy makers perceive the identity of their state will thereafter suggest particular roles, policy objectives and subsequent policy actions. By telling policy makers and strategic planners who they—that is, the state—are and what they are like, strategic culture then shapes the sorts of interests and political goals and outcomes that these policy makers will seek, and strategic planners provide the various strategic instruments and options that presumably allow for these desired goals and outcomes to be realized.[29] It provides the tools with which policy makers and the population can define and understand the situation they are in, interpret adversarial motives, and suggest ways and means by which state interests can be realized despite such adversarial intentions.[30] This parallels role theory of state behavior, which posits that "states are actors that behave consistent with specific roles with which they identify."[31] For instance, it has been argued that the predominantly Caucasian nature of Australian society at least partially explains Australia's historical self-identity as a member of the so-called West, despite its geographical location, and this self-identity made it tilt towards the Australia-New Zealand-United States (ANZUS) alliance, in the first place, and maintain its commitment throughout the 1980s and 1990s to what was by then, arguably, an increasingly anachronistic alliance.[32]

Secondly, and building on the first point, strategic culture shapes how policy makers and strategic planners perceive the international environment, in terms of providing the dominant paradigm through which they understand inter-state relations.[33] This taps into the argument that self-identity and the identification of the 'other' are inextricably inter-linked.[34] Every state exists in, and has to navigate its way through, an increasingly complex strategic environment that comprises a number of potentially competing actors (state as well as non-state) and interests. Whether or not the state exists in a low or high-threat environment, and which of the various actors in its strategic environment comes to be identified as the putative adversary, how these questions are answered is as much a result of rational analysis of 'objective' conditions as it is of cultural perspectives of the 'objective' conditions in the strategic environment. Policy makers use a variety of cultural perspectives to make sense of a vast array of incoming information, most of which will pass by unnoticed. Cultural perspectives from historical experiences, for instance, provide the mechanisms by which policy makers can sift out the unimportant information and concentrate their scarce attention resources on the information that they perceive to be pertinent and relevant to the given situation. Such non-objective considerations thus open up the possibility that policy makers' perceptions and the policy decisions that they make may possibly be discordant with the 'objective' conditions that they face. The process of developing negative stereotypes coincides with the labeling of 'others' as threats on the basis that these 'others' are "trying to prevent the realisation of your national goals," whereas allies are "those who are willing or able to aid you in your attempt to achieve national goals."[35]

In so doing, strategic culture allows policy makers to make assessments about the level and immediacy of the threat posed by the putative adversaries. It is not enough that policy makers perceive their state to exist in a less than completely safe strategic environment; they also need to prioritize these threats, consider how and when specific threats can arise, and the seriousness that they need to accord such specific threats. What is of interest here is the issue of intentions. It is here that the subjective influences, specifically the strategic cultural perspectives, on the process of assessing threats are important.[36] The images and symbols that constitute the state's strategic culture act as a series of filter through which its policy makers and strategic planners make sense of and prioritize the overwhelming amount of information coming from their operating environment. In the search for information, policy makers are necessarily constrained by their ability and willingness to devote limited attention to all sources of information available. Yaacov Vertzberger makes a case for how policy makers give greater attention to incoming information that is generally consonant with prior expectations and beliefs, while discarding what appears to be discordant information.[37] The images and symbols that policy makers use thus form their most basic cognitive tools, and help to influence the assessments policy makers make of developments in the state's strategic environment: for instance, in 1967, in an attempt to reduce growing tensions with their Arab neighbors, Israeli policy makers decided to reduce the military presence in the country's independence day celebrations; except that Arab strategic planners and policy makers interpreted this reduced military presence as evidence of an Israeli Defence Forces (IDF) mobilization elsewhere, and therefore a sign that Israel was planning to initiate military operations against them.

Throughout this discussion, the role of elites—policy makers and strategic planners—has been a central feature of strategic culture theory. Snyder had initially proposed strategic culture as a way to explain Soviet strategic behavior precisely because he was unable to reconcile Soviet strategic behavior with rational choice theory, game theory, and neo-realism, which had over-simplified the state as a unitary and coherent actor. The concept of strategic culture, instead, placed policy makers and strategic planners—the human beings who give the state its almost animate characteristic, and whose policy decisions confer upon the state its identity—front and centre of its analytical focus.[38] If a state has a particular identity, it is the people who populate it that actually confer on the state that self-identity; and in particular, it is the policy makers who enact the policies that confer that particular identity to the state, and their strategic planners whose work supports the policy decisions of the former community.

Except the human community on which strategic culture ought to train its analytical focus is more than just the elites of the state—its strategic planners and policy makers—it should also encompass the wider population: after all, it is this wider population that will transform the policy and strategic choices that these elites make into strategic behavior. The population has to be mobilized into accepting the choices that these elites make, if the national interests of the state—as defined by these elites—are to be secured. Strategic behavior may well reflect the self-interests of these elites, and the 'national interest' be nothing more than the narrow and self-serving interests of these policy elites.[39] Nevertheless, this chapter argues that the population's acceptance, and even embracing, of these policy and strategic decisions suggests that strategic culture has a wider basis of its existence. The identification of narrow elite interests with the wider interests of the state, such that threats to the legitimacy of these elites become conflated with threats to the state, may result in a scenario where, under conditions of domestic political unrest, policy makers often seek to divert domestic political attention to foreign adventures, and seek to bolster their domestic positions through foreign policy successes.[40] However, this is possible only if the wider population happens to concur with their policy elites, that the national interests of the state are being threatened. This is not a purely hypothetical scenario: the Galtieri regime in Argentina was able to mobilize the Argentinian population, in the process diverting the latter's attention away from the regime's economic failures and human rights abuses, by focusing the population's attention on Argentina's territorial dispute with Great Britain over what the Malvinas/Falkland Islands. The images and symbols that constitute a state's strategic culture can thus be manipulated, in other words.

If the images and symbols about a state's identity have a popular subscription, it means strategic culture has to correlate with some form of inference-based logic: it cannot exist independent of an 'empirical' basis, there has to be an apparently 'real' historical basis to these images and symbols. Strategic culture simply does not—indeed, it cannot—exist in a vacuum, but is part of a wider process of socialization that is informed by historical events.[41] The process of national self-identification discussed earlier, the stories of who we are, where we came from, and what we value, the institutional arrangements that we use to organize policy-making processes in terms of both content and structure, these are constructed on a historical foundation.[42] In his book, *Hitler's Willing Executioners*, Daniel Goldhagen argued that both the policy elites and the population were culpable for the development of anti-Semitism in Nazi Germany; the portrayals of Jewish people in Nazi propaganda drew from, correlated with, and then confirmed popularly-held "cosmological and ontological beliefs [that were] well-nigh universal [and seemed] so self-evidently true."[43] The symbols and images that elites use correlate to their own belief systems themselves based on inferences about the operating environment.[44] At the very least, there should not be a complete disjuncture with such popular beliefs; otherwise, it is likely that such elite attempts at political mobilization will fail:

When a conversation is monolithic or close to monolithic, certain points—and this includes the unstated, underlying cognitive models— then a society's members automatically incorporate its features into the organisation of their minds, into the fundamental axioms that they use (consciously or unconsciously) in perceiving, understanding, analysing, and responding to all social phenomena.[45]

In other words, history matters, in the process of perceiving the 'other' as the putative adversary. States do not become hostile to one another because of negative stereotypes; instead, their policy makers, strategic planners and sometimes the population "develop negative stereotypes" about other states and other peoples because their see these other people and states as "preventing them from achieving their self or national interests."[46] South Koreans maintain an ambivalent attitude towards China and Japan, because the latter has historically sought to control the Korean peninsula, thus preventing Koreans from attaining self-determination and independence. However, South Koreans are more to likely see North Korea as an immediate threat for three reasons: it was North Korea's invasion of South Korea that started the Korean War; North Korea has since maintained a more or less hostile stance towards the South; and North Korea possesses nuclear weapons. South Koreans are likely less aware of revisionist Korean War historiography, that demonstrates that the then-President of South Korea, Syngman Rhee, was himself spoiling for a war against the North.[47] South Koreans are also less likely to be aware that the late South Korean dictator, Park Chung Hee, himself contemplated a nuclear weapons programme in the 1970s to compensate for what was thought to be a declining United States security guarantee as a result of the outcome of the United States' war in Vietnam.

Because of the requirement of popular 'buy-in', this means strategic culture is not as malleable as the arguments of Klein and Lucknow, mentioned above, might have otherwise suggested. Strategic cultures can and will change, as the conditions in which states exist evolve over time. Nevertheless, this chapter suggests that strategic culture can also be resistant to change, precisely because of popular subscription to its images and symbols. As Kupchan argued, even if the images and symbols of a state's strategic culture are manipulated by policy elites in an attempt to mobilize the population in pursuit of the former's narrow interests, both groups are susceptible to the process of internalization of these images and symbols.[48]

Tendencies, not Decision—Strategic Culture, Defense Policy and Strategy

The concept of strategic culture appears to connect with, or at least parallel, the 'national ways of war or warfare' body of literature—that there are patterns in the use of force that are unique to specific nations, because different nations have different ways of thinking about the use of force, or because the different states have different material conditions.[49] In the first instance, despite having long-range submarines that were able to operate in the eastern Pacific, and therefore able to interdict sea lines of communication between Hawaii and the western coast of continental United States, the Imperial Japanese Navy continued to deploy these capabilities against the United States Navy, as a part of IJN fleet operations.[50] In the second instance, throughout World War II, the United States Army's preferred method for clearing minefields was to use "profligate artillery barrage" whereas the Soviet Red Army "would expend men (and women) on the same duty."[51]

However, the parallels may be more apparent than actual. In the case of the national styles of war/warfare literature, Basil Liddell Hart's *British Way of Warfare* was more advocacy than analysis and explanation; Russell Weigley's *The American Way of War* relied on the historical method to demonstrate that there was a pattern to the way the United State military thinks about, prepares and conducts war. Snyder developed his theory of strategic culture because he felt that other existing theories of international relations fell short in describing and explaining specific Soviet strategic behavior; Snyder wanted to construct an explanation of the nuances in Soviet strategic behavior that then-conventional literature was not able to. Snyder was careful in his argument to limit the scope and applicability of the concept: as he wrote elsewhere, "culture, including strategic culture, is an explanation to be used only when all else fails."[52]

A criticism of the so-called first-generation of strategic culture theorists has been the element of over-determinism. Colin Gray may have admitted that first-generation theory came close to this charge, but he was oblivious to the subtle self-contradiction in two assertions that appeared on contiguous paragraphs in *Modern Strategy*[53]:

It is important to understand that even when a security community is performing missions that traditionally it has not much favoured, it must behave in a culturally shaped manner. Germans cannot help but be Germans, whether they are waging war as they would prefer or as they must.

In the next paragraph, Gray then asserted:

The claim ... is that culture **shapes** [emphasis mine] the process of strategy-making, and **influences** [emphasis mine] the execution of strategy, no matter how close actual choice may be to some abstract, or idealized cultural preference.

If Germans cannot help but be Germans in their strategic behavior, surely culture determines—not shapes or influences—strategic behavior. If American strategic culture means that the United States military typically does not like a human security-centric approach to counterinsurgency and counterterrorism operations—typically characterized as a 'hearts and minds' approach—that most scholars and strategists insist is necessary for strategic success in those types of scenarios, it fails to account for American strategists and scholars—such as David Petraeus, H.R. McMaster and John Nagl—who have 'bucked the trend' and advocated for a more people-centric approach. Whatever the value proposition of strategic culture to defense policy and strategy may be, over-determinism is clearly a problem that needs to be avoided.

Rather, this chapter argues that the impact that strategic culture exercises on defense policy and strategy is in the realm of strategic preferences and tendencies; what strategic culture is not of strategic decision-making that arise as a result of particular strategic cultures, "a people's distinctive style of dealing with and thinking about the problems of national security."[54]

Strategic culture provides the "prism through which all information pertaining to national security and defence is seen when it is interpreted, judged, and evaluated."[55] In so doing, strategic culture shapes policy-making towards certain preferred paths. When policy makers and strategic planners 'see' putative adversaries in their state's strategic environment, they make assessments about the immediacy (probability) and potential severity (impact) of the threats that these adversaries can pose to the state. The scenarios that occupy the extreme ends of the spectrum—low-probability/low-impact and high-probability/high-impact—are, in a sense, the easiest to address, as the policy responses are obvious. It is the scenarios that occupy the middle of the spectrum where policy responses become more challenging, because of the need to prioritize between the high-probability/low-impact and low-probability/high-impact scenarios: neither choice, arguably, is intuitive and obvious. What is clear is that there ought to be congruence between popular sentiments and policy makers' threat assessments. Policy makers will need to convince the electorate of the former's strategic conceptions in order to justify the expenditure of scarce economic resources on national security requirements. This is where the analogies and symbols through which strategic culture is manifested come in, because policy makers can mobilize these analogies and symbols to shape popular sentiments.

To illustrate, the main policies through which Singapore addresses its national security requirements are conscription and Total Defence.[56] First, conscription of the nation's males of the 18–50 age groups was deemed necessary to generate sufficient human resources to address the multi-faceted nature of Singapore's national security, both domestic and foreign: the foreign threats to national security—the low-probability/high-impact scenario of inter-state war—are the purview of the Ministry of Defence (through the Singapore Armed Forces); whereas the domestic security threats—the high-probability/low-impact scenario of terrorist attacks—are the purview of the Ministry of Home Affairs (through the Singapore Police Force and the Singapore Civil Defence Force).[57] The fact that it is the former that receives lion's share of national security resources—both financial and human—is indicative that Singapore policy makers have placed greater focus on the scenario of inter-state war. Second, Total Defence is the holistic policy through which Singapore's national security actors can address both types of national security problems, nevertheless, it reinforces the argument that Singapore's policy makers have prioritized the scenario of inter-state war, for two reasons. One, Total Defence Day, which is a mechanism through which Singapore's policy makers seek to constantly reinforce the narrative of vulnerability, is held on 15 February of the year, because it was on 15 February 1942 that the British colonial administration in Singapore surrendered to the Imperial Japanese Army, thus ushering in 3 years of hardship and deprivation imposed by the Japanese Occupation. Two, Total Defence Day is understood, even by the Singapore Civil Defence Force, principally through the lens of inter-state war: "The way war is conducted today, our limited resources, the nature of our society and the size of our country, require that our country have a Total Defence capability which involves not only the Singapore Armed Forces (SAF) but also the civilian population."[58]

If strategic culture does not dictate strategic policy, rather, it generates tendencies and preferences towards certain strategic options, these strategic choices will ultimately be manifested in strategic behavior. Colin Gray had a point, in other words, when he insisted that strategic culture influences defense trends, such as force levels and orders of battle, which will impact on strategic behavior in the form of military operations.[59] Or, as Ken Booth stated,

> States … do not necessarily respond to military stimuli and they do not respond in any mechanical way. If they respond at all, they respond

according to national styles and personal idiosyncrasies. The dogs of war, when unleashed, may sometimes exhibit Pavlovian behaviour, but they also have a pedigree.[60]

With regards to the issue of force levels and orders of battle, Elizabeth Kier has argued that, given its economic and political limitations, the logical and sensible policy option for the British Army in the 1920s and 1930s was to transform its hidebound cavalry corps by adopting the then-nascent armored warfare technologies.[61] Except that was precisely what the British Army chose to not do.

Conclusion

This chapter has attempted to develop the argument that strategic culture is not a straitjacket that determines strategic choices by first limiting options, but rather it is a mindset that there are preferred strategic options while always open to other interventions. Strategic culture acts as a set of lenses through which strategic planners and policy makers perceive and make sense of the state's strategic environment. Strategic culture helps these actors to identify the character of the strategic environment as either inherently benign or potentially hostile, and, in the case of the latter, whether or not these putative adversaries will challenge the state's core or peripheral interests. At the same time, strategic culture helps strategic planners and policy makers to assess, whether correctly or incorrectly, the immediacy of these challenges to the state's interests. Furthermore, the analogies, images and symbols through which this strategic environment is understood, because it is either based on or parallels historical experiences, allow policy makers to subsequently mobilize public opinion such that strategic actions, when necessary, are possible. This chapter therefore suggests an understanding of strategic culture that avoids the deterministic tendencies in first-generation strategic culture theory, and moves closer towards an essentially elite-level analysis of strategic culture.

    Strategic culture by itself cannot adequately explain the threat perceptions of the strategic planners and policy makers of the state, of course. Strategic culture is about how these actors perceive the environment in which the state exists, and because it operates in the realm of perceptions and mindsets, constitutes only one half of the equation. Equally important is the actual environment in which the state exists. This is the geography in which the state finds itself. This is the focus of the next chapter: the physical reality of this environment, specifically, political and strategic geography.

Endnotes

[1] Strategic culture is, admittedly, a controversial topic. For a useful summary of these controversies, see: Lawrence Sondhaus, *Strategic Culture and Ways of War: An Historical Overview* (London and New York: Routledge, 2006), pp. 1–14.

[2] Colin S. Gray, *Modern Strategy* (Oxford: Oxford University Press, 1999), pp. 129–151. Also, see: Yuen Foong Khong, *Analogies At War: Korea, Munich, Dien Bien Phu, and the Vietnam Decisions of 1965* (Oxford and Princeton: Princeton University Press, 1992); Jack L. Snyder, *The Ideology of the Offensive: Military Decision-Making and the Disasters of 1914* (Ithaca & London: Cornell University Press, 1984).

[3] Richard Ned Lebow, *Between Peace and War: The Nature of International Crisis* (Baltimore: Johns Hopkins University Press, 1981), p. 229. Also see: Eric Herring, *Danger and Opportunity: Explaining International Crises Outcomes* (Manchester and New York: Manchester University Press, 1995), p. 70.

[4] Colin S. Gray, *The Geopolitics of Superpower* (Lexington: University Press of Kentucky, 1988), pp. 42–43.

[5] Charles A. Kupchan, *The Vulnerability of Empire* (Ithaca and London: Cornell University Press, 1994), pp. 26–30.

[6] Stephen Peter Rosen, "Military Effectiveness: Why Society Matters," *International Security*, vol. 19, no. 4, 1995, pp. 14–17. Tyler's famous definition of culture posits culture as "that complex whole which includes knowledge, belief, art, law, morals, custom, and any other capabilities and habits acquired by man as a member of society." Inherent in Tyler's conceptualization of culture was the argument that all human cultures evolve through stages of development, from barbarism to a pinnacle of rational science and secularism. See: Edward Burnett Tyler, *Primitive Culture* (New York: J.P. Putnam's Sons, 1920), p. 1. Tyler's conception of culture as a progression towards a singular pinnacle was challenged by the German anthropologist, Franz Boas, who developed the cultural relativism argument, that there are different cultures, because these differences are rooted in different contexts, different histories. For a succinct discussion of Boas, see: Adam Kuper, *Culture: The Anthropologists' Account* (Cambridge, Massachusetts: Harvard University Press, 1999), p. 56.

[7] Colin S. Gray, "Inescapable Geography," in Colin S. Gray and Geoffrey Sloan (eds.), *Geopolitics, Geography and Strategy* (London and Portland: Frank Cass, 1999). p. 140.

[8] Kupchan, *The Vulnerability of Empire*, p. 89.

[9] *Ibid.*, p. 29.

[10] Jack L. Snyder, *The Soviet Strategic Culture: Implications for Limited Nuclear Operations*, RAND R-2154-AF (Santa Monica: RAND, 1977), p. 38.

[11] Jack L. Snyder, "The Concept of Strategic Culture: Caveat Emptor," in Carl G. Jacobsen (ed.), *Strategic Power: USA/USSR* (Basingstoke and London: Macmillan, 1990), pp. 3–9.

[12] Colin S. Gray, "National Style in Strategy: The American Example," *International Security*, vol. 6, no. 2, 1981, p. 22.

[13] Gray, *Modern Strategy*, pp. 132–136.

[14] Ken Booth, *Strategy and Ethnocentrism* (New York: Holmes and Meier, 1979), pp. 14–15.

[15] *Ibid.*, p. 16.

[16] Ken Booth, "The Concept of Strategic Culture Confirmed," in Carl G. Jacobsen (ed.), *Strategic Power: USA/USSR* (Basingstoke and London: Macmillan, 1990), p. 121.

[17] Alistair Iain Johnston, "Thinking About Strategic Culture," *International Security*, vol. 19, no. 4, 1995, p. 63.

[18] For second-generation theory, see, for instance: Bradley S. Klein, "Hegemony and Strategic Culture: American Power Projection and Alliance Defence Policies," *Review of International Studies*, vol. 14, no. 2, 1988, pp. 133–148; and, Robin Luckham, "Armament Culture," *Alternatives*, vol. 10, no. 10, 1984, pp. 1–44.

[19] Alistair Iain Johnston, *Cultural Realism: Strategic Culture and Grand Strategy in Chinese History* (Princeton: Princeton University Press, 1995), pp. 35–36.

[20] Alan Bloomfield, "Time to Move on: Reconceptualizing the Strategic Culture Debate," *Contemporary Security Policy*, vol. 33, no. 3, 2012, p. 450.

[21] Gray, *Modern Strategy*, pp. 134–136.

[22] *Ibid.*, pp. 132–133.

[23] *Ibid.*, pp. 135–136.

[24] Lawrence Sondhaus, *Strategic Culture and Ways of War* (London and New York: Routledge, 2006), p. 12.

[25] *Ibid.*, p. 9.

[26] See, for instance: David G. Haglund, "What good is strategic culture? A modest defence of an immodest concept," *International Journal*, vol. 59, no. 3, 2004, pp. 479–501; Stuart Poore, "What is the Context? A Reply to the Gray-Johnston Debate on Strategic Culture," *Review of International Studies*, vol. 29, no. 2, 2003, pp. 279–284.

[27] Bloomfield, "Time to Move On," pp. 451–451. Also, see: Bernard Fook Weng Loo, *Middle Powers and Accidental Wars: A Study in Conventional Strategic Stability* (Lewiston, NY: Edwin Mellen Press, 2005), pp. 35–68.

[28] See, for instance: Derek da Cunha, "Defence and Security: Evolving Threat Perceptions," in Derek da Cunha (ed.), *Singapore in the New Millennium: Challenges Facing the City-State* (Singapore: Institute of Southeast Asian

Studies, 2002), pp. 133–153; Chong Guan Kwa, "Relating to the World: Images, Metaphors, and Analogies," in Derek da Cunha (ed.), *Singapore in the New Millennium: Challenges Facing the City-State* (Singapore: Institute of Southeast Asian Studies, 2002), pp. 108–132; Bernard F.W. Loo, "From Poisoned Shrimp to Porcupine to Dolphin: Cultural and Geographic Perspectives of the Evolution of Singapore's Strategic Posture," in Amitav Acharya and Lee Lai To (eds.), *Asia in the New Millennium: APISA First Congress Proceedings, 27–30 November 2003* (Singapore: Marshall Cavendish Academic, 2004), pp. 352–375; Bilveer Singh, *The Vulnerability of Small States Revisited: A Study of Singapore's Post-Cold War Foreign Policy* (Yogyakarta: Gajah Mada University Press, 1999).

[29]Valerie M. Hudson, "Culture and Foreign Policy: Developing a Research Agenda," in Valerie M. Hudson (ed.), *Culture and Foreign Policy* (Boulder and London: Lynne Rienner Publishers, 1997), pp: 1–26; Ronald L. Jepperson, Alexander Wendt, and Peter J. Katzenstein Norms, "Identity, and Culture in National Security," in Peter J. Katzenstein (ed.), *The Culture of National Security: Norms and Identity in World Politics* (New York: Columbia University Press, 1996) pp. 54–65.

[30]Sanjoy Banerjee, "The Cultural Logic of National Identity Formation: Contending Discourses in Late Colonial India," in Valerie M. Hudson (ed.), *Culture and Foreign Policy* (Boulder and London: Lynne Rienner Publishers, 1997), p. 7.

[31]Glenn Chafetz, Hillel Abramson, and Suzette Grillot, "Culture and National Role Conceptions: Belorussian and Ukrainian Compliance with the Nuclear Non-Proliferation Regime," in Valerie M. Hudson (ed.), *Culture and Foreign Policy* (Boulder and London: Lynne Rienner Publishers, 1997), p. 174.

[32]Simon Dalby, "Security Discourse, the ANZUS Alliance and Australian Identity," in Graeme Cheeseman and Robert Bruce (eds.), *Discourses of Danger & Dread Frontiers: Australian Defence and Security Thinking after the Cold War* (St Leonards, NSW: Allen & Unwin Australia Pty. Ltd., 1996), pp. 108–133.

[33]Kupchan, *The Vulnerability of Empire*, p. 41.

[34]Michael N. Barnett, "Identity and Alliances in the Middle East," in Peter J. Katzenstein (ed.), *The Culture of National Security: Norms and Identity in World Politics* (New York: Columbia University Press, 1996), pp. 407–408.

[35]Asher Arian, *Security Threatened: Surveying Israeli Opinion on Peace and War* (Tel Aviv: Jaffee Center for Strategic Studies, Tel Aviv University; Cambridge and New York: Cambridge University Press, 1995), p. 24.

[36]Colin S. Gray, *Weapons Don't Make War: Policy, Strategy, & Military Technology* (Lawrence, Kansas: University Press of Kansas, 1993), pp. 90–91.

[37]Yaacov Y.I. Vertzberger, *The World In Their Minds: Information Processing, Cognition, and Perception in Foreign Policy Decisionmaking* (Stanford, California: Stanford University Press, 1990), pp. 50–144.

[38]Snyder, *The Soviet Strategic Culture*, p. 40.

[39] This idea that strategic choices reflect elite self-interests, rather than something ethereal like strategic culture, can be found in, for instance: Bradley S. Klein, "Hegemony and Strategic Culture: American Power Projection and Alliance Defence Politics," *Review of International Studies*, vol. 14, no. 2, 1988, pp. 133–148; Robin Luckham, "Armament Culture," *Alternatives*, no. 10, 1984, pp. 1–44.

[40] Kupchan, *The Vulnerability of Empire*, p. 20.

[41] David T. Twining, "Soviet Strategic Culture—The Missing Dimension," *Intelligence and National Security*, vol. 4, no. 1, 1989, p. 177.

[42] Marijke Breuning, "Culture, History, Role: Belgian and Dutch Axioms and Foreign Assistance Policy," in Valerie M. Hudson (ed.), *Culture and Foreign Policy* (Boulder and London: Lynne Rienner Publishers, 1997), p. 101.

[43] Daniel Johan Goldhagen, *Hitler's Willing Executioners: Ordinary Germans and the Holocaust* (London: Abacus, 1996), pp. 28–29.

[44] Kupchan, *The Vulnerability of Empire*, p. 41.

[45] Goldhagen, *Hitler's Willing Executioners*, pp. 33–34.

[46] Arian, *Security Threatened*, pp. 24–25.

[47] See, in particular: Bruce Cumings, *The Origins of the Korean War vol. I: Liberation and the Emergence of Separate Regimes, 1945–1947* (Princeton: Princeton University Press, 1981); Bruce Cumings, *The Origins of the Korean War vol. II: The Roaring of the Cataract, 1947–1950* (Princeton: Prince ton University Press, 1990). Also: John Merrill, *Korea: The Peninsular Origins of the War* (Newark: University of Delaware Press, 1989); Robert Simmons, *The Strained Alliance: Peking, Pyongyang, Moscow and the Politics of the Korean Civil War* (New York: Free Press, 1975).

[48] Kupchan, *The Vulnerability of Empire*, pp. 23–24.

[49] See, for instance: Basil Henry Liddell Hart, *The British Way in Warfare* (London: Faber & Faber, 1932); Russell Weigley, *The American Way of War: A History of United States Military Strategy and Policy* (New York: Macmillan, 1973). For a recent discussion of the British way of warfare, see: Colin McInnes, *Hot War, Cold War: The British Army's Way in Warfare, 1945–95* (London: Brassey's, 1996).

[50] W. Clayton James, "American and Japanese Strategies in the Pacific War," in Peter Paret, Gordon A. Craig and Felix Gilbert (eds.), *Makers of Modern Strategy from Machiavelli to the Nuclear Age* (Princeton, New Jersey: Princeton University Press, 1986), p. 707.

[51] Colin S. Gray, "National Style in Strategy: The American Example," *International Security*, vol. 6, no. 2, 1981, p. 38.

[52] Cited in: Sondhaus, *Strategic Culture and Ways of War*, p. 4.

[53] Gray, *Modern Strategy*, p. 135.

[54] Rosita Dellious, "'How May the World Be at Peace?' Idealism as Realism in Chinese Strategic Culture," in Valerie M. Hudson (ed.), *Culture and Foreign Policy* (Boulder and London: Lynne Rienner Publishers, 1997), p. 202.
[55] Vertzberger, *The World In Their Minds*, p. 272.
[56] Bernard F.W. Loo, "Goh Keng Swee and the Emergence of a Modern SAF: The Rearing of a Poisonous Shrimp," in Emrys Chew and Kwa Chong Guan (eds.), *Goh Keng Swee: A Legacy of Public Service* (Singapore: World Scientific, 2012), pp. 127–182.
[57] For an in-depth analysis of conscription and the Singapore Armed Forces, see: Bernard F.W. Loo, "Conscription and its Contribution to Singapore," in S. Vasoo and Bilveer Singh (eds.), *Critical Issues in Asset Building in Singapore's Development* (Singapore: World Scientific, 2018), pp. 149–162.
[58] https://www.scdf.gov.sg/home/community-volunteers/community-preparedness/total-defence.
[59] Colin S. Gray, *Weapons Don't Make War: Policy, Strategy, & Military Technology* (Lawrence, Kansas: University Press of Kansas, 1993), pp. 90–91.
[60] Booth, *Strategy and Ethnocentrism*, p. 23.
[61] Elizabeth Kier, "Culture and Military Doctrine," *International Security*, vol. 19, no. 4, 1995, p. 83.

Chapter 3

Singapore's Military History 2.0

Weichong Ong*

*Assistant Professor, National Security Studies Programme,
S Rajaratnam School of Internal Studies,
Nanyang Technological University, Singapore*

iswcong@ntu.edu.sg

The eminent historian, Wang Gungwu, noted that in most contemporary Southeast Asian countries, historians are obliged to "contribute to nation-building efforts by writing national history."[1] Military history and those who write it are no different. The Southeast Asian experience of World War II (WWII) marked by the defeat of Western colonial powers at the hands of Japan "served as both a catalyst and an inspiration to Southeast Asian nationalist movements."[2] In much of post-war Southeast Asia, the central role played by militaries in the struggle for independence such as the Tentara Nasional Indonesia (TNI) in Indonesia earned the armed forces a place in the nation-building narrative. In the case of Vietnam, the narrative of armed resistance pre-dates the founding of modern nation-states in Asia. Vietnamese nationalism draws from a common strand of armed resistance from ancient times to the modern era—that dates from its resistance to Chinese domination from the Han Dynasty, French rule

---

*An earlier version of this chapter was published in: Tristan Moss and Rom Richardson (eds.), *Beyond Combat: Australian Military Activity Away From the Battlefields* (Sydney: UNSW Press, 2018). The author thanks NewSouth and UNSW Press for their kind permission to have it reprinted here.

from the 19th century and American intervention during the Cold War. Unlike Vietnam, Singapore does not have the luxury of a usable pre-modern, pre-colonial military past. Neither can Singapore invoke the martial spirit of an anti-colonial revolutionary past like Indonesia. In the absence of a war of armed resistance to colonial authority, it is perhaps understandable that in order for Singapore to construct its military history, drawing upon its involvement in British colonial rule is a convenient starting point.

The focus on the common theme of shared wartime suffering allows for Singapore's pre-independence military past—particularly events, sites, monuments and personages from WWII—to be appropriated as part of Singapore's nation-building narrative. Sites that represent such shared memories, such as the Kranji War Cemetery, are used and remembered by different groups in different ways. Whilst the Australians meet at Kranji on ANZAC Day to commemorate the manifestation of the ANZAC Spirit, the Kranji war graves stand as testament to British stoicism and the notion of "the captive as hero."[3] In contrast, Singaporean school children and Full-time National Servicemen (NSFs) visit the site to show respect for the fallen and learn the lesson of what may come if Singapore were to rely on any external power for its national security. The sheer scale and number of parties involved in the Battle and Fall of Singapore (Australian, British, Chinese, Indian, Malay and Japanese) have subjected the event to the (re)interpretations of each group "according to their own rituals and needs."[4] In short, colonial war memory and remembrance co-exists alongside commemoration and nation-building lessons for an emergent nation-state.

Such appropriations of colonial history into Singapore's 'nation-building' story can be problematic. The logic of colonial defense demands that "colonial regimes had to been defended by others. If they had to defend themselves, they could no longer be colonial."[5] The two main WWII figures appropriated as Singapore's war heroes, Lim Bo Seng and Adnan Bin Saidi, fought for ideals 'other' than Singapore's independence. Lim Bo Seng, a Nationalist Chinese Colonel and a British Force 136 officer (established by British Special Operations Executive to support the various resistance movements in Japanese-occupied Southeast Asia) was a war hero of the Overseas Chinese (*hua qiao*) who, like many his DALFORCE (or, the Singapore Overseas Chinese Volunteer Army) comrades, saw a *Kuomingtang* (KMT) led China as their motherland (*zuguo*).[6] DALFORCE, a volunteer militia drawn from both KMT supporters and communists just before the fall of Singapore, was "a symbol of Overseas Chinese unity, resistance and identity."[7] There were cracks however behind this 'united front' of Overseas Chinese military involvement. Despite their common cause against the Japanese, unlike the non-communist members of Force 136, Overseas Chinese with communist leanings who joined or supported Chin Peng's communist-led Malaya Peoples' Anti-Japanese Army (MPAJA) were treated with suspicion by Allied commanders in Southeast Asia Command (SEAC).[8] This appropriation of Lim Bo Seng as a Singapore War Hero presents a problem. To a younger generation of Singaporeans born after independence or 'Post-65ers' who may not necessarily identify with the values of Singapore's *hua qiao*, "Lim's heroism is beyond reproach, but his actual nationality and [intentions] make his appropriateness in the Singapore Story debatable."[9]

For the Malay community, the heroic stand made by the Malay Regiment and by Adnan at Pasir Panjang Ridge at the height of the Battle for Singapore represents the "distinctive martial qualities that boys in the emerging nation-state should emulate," but this same battle is represented in Malaysia as part of a Malay martial tradition that dates back to Hang Tuah (a legendary *Laksamana*, or admiral of the Malacca Sultanate) and a manifestation of Malay nationalism.[10] The appropriations of Lim Bo Seng and Adnan as Singapore war heroes are perhaps understandable if military heroes are strictly defined by martial virtues and sacrifice demonstrated in war and battle. There are however Singaporean military figures and a distinctive brand of Singapore military history if the parameters are broadened beyond the traditional 'drums and trumpet' approach.

Post-colonial Singapore is not devoid of a military history, but it is one that is largely shaped by the experiences of its peacetime citizen military. When Singapore became an independent state responsible for its own defense, this was something neither British nor local leaders planned or intended.[11] The end of empire, the precipitated withdrawal of British forces and the separation of Singapore from the Malaysian federation meant that the 'accidental nation' of Singapore had to create armed forces virtually from scratch. The nascent state of Singapore decided that the best way to build up a credible defense force was to create a citizen army with the help of Israeli advisors. The rationale is best explained in Singapore's Founding Prime Minister, Lee Kuan Yew's own words: "This was an ambitious plan based on the Israeli practice of mobilizing the maximum number possible in the shortest time possible. We thought it important for people in and outside Singapore to know that despite our small population, we could mobilize a large fighting force at short notice."[12]

The legacy of Lee Kuan Yew's 'Third World to First' narrative, or 'Singapore Story 1.0' shapes the core of Singapore's post-colonial military history. This story is a distinct departure from colonial Singapore's 'drums and trumpets' past of WWII. Rather than tales of sacrifice in pitched battles, it is from the collective memory of the Singapore Armed Force's (SAF) peacetime experiences both at home and in overseas deployments that a distinct post-colonial Singaporean military history has emerged. By moving beyond the battlefield into the 'interface between war and society' and the social composition of the SAF, a 'new military history' of Singapore that is Singaporean can be written.[13] This history has been written in two phases. Despite the hegemonic influence of 'Singapore Story 1.0', more pluralistic strands are finding a voice in the crafting of Singapore's military history—a Singapore Military History 2.0. This chapter explains how despite the absence of combat experience, a distinct brand of military history focused on the peacetime development of the SAF buttressed by the memories and voices of its citizen soldiers is beginning to emerge. In addition, the actual use and perceived utility of military history within Singapore's professional military education and training (PMET) and the field of Strategic Studies is examined.

Singapore Story 1.0: The Vulnerability-Survival Strand

The 'city-under-siege' or vulnerability part of Singapore Story 1.0 can be traced back to memories, events and decisions of being let down by British Imperial defense during the colonial era. Modern Singapore's sense of insecurity stems primarily from its structural vulnerabilities in geographical location, small size (in land mass, population and natural resources) and historical memories that appear to vindicate perceptions of its vulnerability. The day that Singapore fell to the Japanese, 15 February 1942, has been observed as 'Total Defence Day' by schools in Singapore since 1998. '15 February' is a central part of Singapore Story 1.0 for it conjures from the past the heavy price of occupation should Singapore ever again fall to external aggression. The precipitated withdrawal of British Military forces from Singapore in 1971 against the backdrop of a resurgent Communist Party of Malaya (CPM) insurgency in neighboring Malaysia and increased CPM subversion in Singapore further enhanced the climate of insecurity in the new state.

This sense of vulnerability is further accentuated by Singapore's geopolitical position as a predominantly ethnic Chinese city-state nestled between two much larger Malay-Muslim neighbors. At independence, there were concerns that "Malaysia would seek to dominate Singapore using its substantial military capabilities."[14] Despite much improved bilateral relations since independence, reminders of Singapore's smallness were signaled from time to time by its immediate neighbors. Former Indonesian President B.J. Habibie's derogatory reference to Singapore as a "Little Red Dot" in 1998 has not only stuck, but became the adopted title of a book on Singapore's Foreign Service and diplomacy.[15] In 2002, strains in bilateral relations emanating from Singapore's need to grow its physical space through land reclamation sparked off an editorial in the Kuala Lumpur-based *New Straits Times* which described Singapore as an "irritating pimple which refuses to burst."[16]

Singapore's near-complete reliance on overseas sources for basic needs such as food, fuel and a large part of its water supply is a major vulnerability.[17] Access to the high seas on which Singapore relies for most of its trade and import of its existential needs must pass through its neighbors' territorial waters. As Tim Huxley points out, any disruption of Singapore's maritime lifelines "would threaten not just its economic well-being—its very survival as an independent nation would be at stake."[18] Even in the absence of a clearly defined enemy, the strategic vulnerabilities inherent in Singapore's geostrategic position, physical limitations and lack of strategic depth provide the main basis for strategic planning, defense doctrine, strategic posture and milieu for the vulnerability-survival strand of Singapore Story 1.0.

The vulnerability-survival narrative of Singapore Story 1.0 on which independent Singapore's military history is built is distinct from most other Southeast countries. Unlike the nation-building narratives of Indonesia and Vietnam where military force or institutions played a central role in the struggle for independence, the SAF story is subsumed within the larger nation-building story. Singapore's first Defence Minister, Goh Keng Swee recognized that the "defence of small states had to be approached in a particular way."[19] For its military defense, Singapore looked to Israel rather than the United Kingdom for its first military advisors in 1965—another emergent state in similar geostrategic circumstances that has survived against the odds. The first Israeli military advisory team arrived in late 1965 and the mission was to remain in Singapore until 1974.[20]

Under the guidance of Israeli military advisors from the mid-60s to the early 70s, the SAF's organization, doctrine, training and equipment were developed "to provide the foundations for the only strategy which made sense"—"deterrence through pre-emption" within Singapore's immediate region.[21] By the 1970s, the twin policies of conscription and sustained military spending provided Singapore with a sizable force of some 300,000 regular, conscript and reserve forces "to pose a reasonable deterrence towards Singapore's two historically-hostile neighbours."[22] This period witnessed the build-up of a basic credible defense force, the 'First Generation' SAF. From the 1980s to the early 1990s, the SAF acquired the capability of limited regional power projection—particularly the protection of Singapore's vital maritime lifelines up to 1,000 miles from home in the event of any regional conflict.[23] The SAF thus made the transition from a 'First Generation' basic defense force to a 'Second Generation' force capable of forward defense within its immediate region. This particular narrative of building a credible military defense against the odds and careful investment in long-term strategic planning is incorporated as part of the larger Singapore Story 1.0. It is also part of a distinct Singaporean military history that tells the story of building the 'First' to 'Second' Generation SAF from the 1960s to 1990s.

Beyond the Vulnerability-Survival Narrative

The SAF's role and capabilities have evolved from 'Rising to the Defence of Singapore' in the 1960s to a more 'Global, More Capable and Ready' one in the 1990s. The more sizable deployments of the SAF since the 1990s reflect a greater confidence in its operational capabilities and desire to develop greater interoperability with multinational partners in coalition missions.[24] The notion of an 'Always Ready' citizen soldier capable of overseas deployment is an operational reality, but the readiness of Singaporean society to accept and risk the frequent and extended deployment of its citizen soldiers in international missions is another matter. After more than 50 years of nation-building, the role and acceptance of conscription, or better known as National Service (NS) has become embedded in the national psyche and social fabric of Singapore. This acceptance however is premised on the defense of Singapore and grounded in the implicit trust that the state will not risk the lives and well-being of its citizen soldiers unnecessarily.

A nation-building narrative focused on the 'vulnerability-survival' strand reflects the enduring structural vulnerabilities of modern Singapore. There is however a cognizance, even amongst establishment historians that the Singapore Story needs to be pluralized "beyond the contributions of Lee Kuan Yew and his PAP Old Guard."[25] One such additional strand is to set Singapore's history in the regional and global context. Historically, Singapore's central place between Southeast Asia and the wider world is "a carefully constructed, and oft-reconstructed, artifice."[26] This also can be said of Singapore's military history. In the colonial era, Singapore was constructed as a symbol of Pax Britannica's wide-reaching presence and control in the Far East.[27] The military history of the 'Third' Generation SAF from the 2000s to present day is currently being written. In perspective, it presents a larger regional and global picture than that from the 1960s-90s. Indeed, Singapore's strategic outlook in the twenty-first century is best summarized in a defense policy paper published in 2000:

> The Asian economic crisis has demonstrated how closely intertwined the interests of nations have become in a borderless world. A small and open country like Singapore is especially susceptible to unpredictable shifts in the international environment. This vulnerability will increase as we become more integrated with the global economy. What happens in another part of the world can have immediate and great spill-over effects on our economy and security. But we cannot turn back from globalisation ... We will have to work more actively with others to safe-guard peace and stability in the region and beyond, to promote a peaceful environment conducive to socio-economic development.[28]

It is of no coincidence that the larger of Singapore's overseas deployments in the first half of the twenty-first century have been in Maritime Security Operations (MSO) in the North Arabian Gulf (Combined Task Force-158) and the Gulf of Aden (Combined Task Force-151). Any potential threat to the global sea-based trade system directly endangers Singapore's security and position as a global financial and maritime hub. Singapore's maritime strategic history can therefore be seen as "a defense of the trading system against the instabilities and conflicts ashore that might threaten it."[29] Rather than using the 'founding' of modern

Singapore by the British in 1819 as the starting point of Singapore's maritime history, there is in recent years, an increasing effort to situate Singapore's maritime hub role within the *Longue Duree* of how various Southeast Asian ports 'have struggled for [regional] supremacy' in the last 2,000 years.[30] John Miksic makes the case that in pre-British Singapore, the strategic importance of Singapore in the Malayu-Jambi (1025–1275), Classical Singapore (1275–1400), Melaka (1400–1511), Johore-Riau (1511–1780) and Dutch-Bugis (1700–1819) phases largely rose and fell with external trends and perceptions.[31] This approach grapples with the geostrategic importance of Singapore or "the influence of military considerations" on the development of Singapore over centuries from Pre-British to modern Singapore.[32]

Despite a significant built-up of its force projection capabilities over the past 50 years, the SAF has yet to deploy any of its forces in overseas combat missions. To date, the deployment and employment of SAF troops and assets in international missions have been in reconstruction, combat service support, humanitarian assistance and disaster relief (HADR) and constabulary roles. From 2003 to 2008, Singapore's contributions to Multinational Force Iraq (MNF-I) included deployments of C-130 transport aircraft, C-135 tanker aircraft and naval assets in the North Arabian Gulf.[33] From 2007 to 2013, Singapore's presence in ISAF Afghanistan included a construction engineering/medical team (which served as part of the New Zealand PRT in Bamiyan), a Weapon Locating Radar (WLR) team in Tarin Kowt, Oruzgan, institutional trainers, intelligence analysts, an Unmanned Aerial Vehicle team (UAV) Task Group, and KC-135 tanker aircraft.[34] The SAF's deployments in CTF-158 and CTF-151 particularly involved the heavy use of unmanned technology and ship-borne automation in MSO. SAF deployments in overseas missions tend to be in selected low-risk non-combat 'niche' roles—particularly where technology can comfortably mitigate the lack of 'boots on the ground' and fulfil its full potential as a force-multiplier.[35] The stories of these deployments are increasingly being told in commemorative books commissioned by the Ministry of Defence such as *2263 Days Operation Blue Ridge: The SAF's Six-year Mission in Afghanistan*, but they are not official histories in British, American or Australian tradition.[36]

Operation Flying Eagle (OFE), the SAF's deployment to the tsunami-hit Indonesian province of Aceh in 2005 and its largest HADR operation to date proved to be an invaluable experience for the SAF and a limited test of its deepened capabilities that have been developed since the

1990s.[37] OFE presents a teachable case of how the SAF's knowledge of local Indonesian culture, politics and language could be put to good use. OFE also witnessed a significant deployment of the SAF's Rapid Deployment Force from the Army's Guards formation.[38] Moreover, despite the ups and downs of Indo-Singaporean state diplomatic relations, OFE tells the story of how existing links between the SAF and the TNI have been strengthened and new ones forged. Similar to the SAF's Afghanistan deployment, a commemorative book *Reaching Out: Operation Flying Eagle* was commissioned by the Ministry of Defence, but written by a defense correspondent of the Straits Times.[39] *Reaching Out: Operation Flying Eagle* provides more operational details and personal stories than *Operation Blue Ridge: The SAF's Six-year Mission in Afghanistan*, but again it falls short of the criteria of an official military history.

Since its independence, Singapore has not been involved in any wars, but that does not mean that there is an absence of a military identity that binds. NS is an accepted fixture of Singaporean life and is seen as a rite of passage. It is underpinned not only by a social contract between Singaporeans and the state, but also the SAF's role in shaping Singaporean society. For the past fifty-odd years, the NS story is told as a 'necessity' for the continued prosperity of Singapore and its citizens, but it is also the social history of a common shared experience that transcends and binds Singaporeans of different ethnic, social and generational backgrounds. This narrative is strongly reflected in the taglines that commemorate recent milestones of NS as a Singaporean institution. 'From Father to Sons', 'Every Singaporean Son' 'Generation to Generation' and 'From my Generation to Yours' are some of the examples that headline the narrative in NS commemoration efforts since 2012. The enduring continuity of NS as a Singaporean institution is emphasized and commemorated in a young nation state that is only beginning to define its national identity. In fact, there are stories of 'everyday' Singaporean military heroes that can form the basis of Singapore's post-1965 collective military memory.

In 2015, Singapore and the SAF celebrated its fiftieth birthday. In *Giving Strength to our Nation: The SAF and Its People*, a book that commemorates the SAF's 50 anniversary, literary critic and poet Gwee Li Sui noted that through NS poems, short stories and plays, "the military world becomes an indirect means to depict and comment on society at large."[40] Likewise, portraying "army life as part of the Singaporean landscape" and capturing the "subtler, private memories and responses to the experience of being in the [SAF]" has been an increasingly fruitful theme for Singaporean art since the 1990s.[41] The SAF is not an all professional military divorced from society, but a citizen force that mirrors Singaporean society at large. Beyond the state sanctioned narratives, Singapore's military history 2.0 is increasingly pluralized by the personalized depictions of its citizen soldiers that recount and remember a shared formative experience. Hence, a shift beyond the battlefield into the interface between the SAF and Singapore society reveals a 'new military history' of Singapore that is distinct from Singapore's colonial military history.

A Usable Past in Singapore's PMET and Strategic Studies

The writing of military history, however, can be distinct from a usable past in PMET or simply put, the idea that military history is of use to the military professional. In his 1961 RUSI lecture "The Use and Abuse of Military History," Sir Michael Howard made the case that for the competent military professional, military history should be pursued in width, in depth and in context.[42] The context of his lecture is important as it specifically addressed academic concerns of military history being used as a the tool of 'myth-making'. Like Plato, Howard believed that myths play a useful social function in preparing young soldiers for the realities of war—a 'nursery history' approach to be measured by a different standard.[43] The creation of a romanticized image of the past in 'myth-making' to inspire and sustain action is not an abuse of history, but a different function from the critical examination of military history by an academic historian or strategic thinker. Thus, it is important to make a distinction in the different approaches that military history is used and its perceived utility in PMET and Strategic Studies.

In the case of PMET, Singapore's WWII history is perceived as a usable past that can be used to draw out stories and lessons of leadership for the SAF. Battlefield tours and lectures on the Battle and Fall of Singapore are incorporated into the leadership training programmes of officer cadets and junior officers. Central to this approach are the stories of wartime resistance and sacrifice that seek to connect the military values of the past to a present generation of military leaders that have not experienced war. This narrow focus that relies on metanarratives of wartime heroism and in certain cases romanticism as inspiration however is differ-ent from using military history to develop what Elliot Cohen describes as the "historical mind"—"a way of thinking that uses history as a mode of inquiry and a framework for thinking about problems."[44]

The 'historical mind' approach to military history is a utilitarian one that looks to the history of war and warfare as a guide to decision making. Rather than look to the past for myths that connect and inspire in junior leadership training, the "historical mind" is employed to educate strategic thinkers in the SAF at the mid-career level. Instead of being fed with stories of wartime heroism from Singapore's WWII past, students of strategic studies at the Goh Keng Swee Command and Staff College approach the same case study as an exercise in contextual deci-sion-making that puts them in the shoes of key decision-makers who are required to examine the critical 'what', 'how' and 'why' questions. This approach lies at the heart of the Clausewitzian approach to military his-tory. To use history as a guide to judgement that enables the military leader to refight campaigns and battles in all their ambiguity, contingency, danger and complexity.[45] when students and makers of strategy look to historical examples to illuminate current predicaments, this is where the 'historical mind' meets the "strategic mind."[46] It offers the student strate-gist the prospects of "wisdom without tears," insights without the costly sacrifice that a battlefield education might require.[47] The context for illumination, however, should be broadened to embrace the complexity of conflict in its entirety rather than a laser focus on major combat operations—which are critical but not always decisive in winning wars or building an enduring peace.

Looking to military history for a usable past to enhance strategic judgement is essential for the military professional. Its utility however can be limited when historical examples are misused. On the potential to use and abuse military history, Clausewitz warned: "Historical examples clarify everything and also provide the best kind of proof in the empirical sciences. This is particularly true of the art of war ... Historical examples are, however, seldom used to such good effect ... To teach the art of war entirely by historical examples ... would be an achievement of the utmost value; but it would be more than the work of lifetime."[48] Just as the writing and interpretation of Singapore's military history is an ongoing process, it is crucial that Singapore's military professionals embrace Clausewitz's exhortation of life-long learning in the shaping and educa-tion of their historical and strategic minds.

Conclusion

Structural vulnerabilities further amplified by deeply rooted historical memories continue to inform much of Singapore's national narrative. The core of Singapore story 1.0 however is increasingly being supplemented by other attendant strands that can co-exist alongside the 'vulnerability-survival' strand. This plurality is also reflected in the sub-field of Singapore's military history. What Singapore lacks in a usable pre-modern, pre-colonial, or anti-colonial revolutionary military past does not mean an absence of a Singaporean military history. Despite the absence of combat experience, a distinct brand of military history focused on the peacetime development of the SAF buttressed by the memories and voices of its citizen soldiers is beginning to emerge in the scripting of a Singapore military history 2.0.

The (re)interpretation of Singapore's WWII past, colonial war memory and remembrance will continue to co-exist alongside commemoration and nation-building lessons for Singapore the nation-state. However, appropriations of colonial history into Singapore's 'nation-building' story can be problematic. A move beyond the battlefield into the interface between the SAF and Singapore society however reveals a 'new military history' of Singapore that is more accessible. In the future, rather than simply appropriate war heroes from Singapore's colonial and pre-independence past, Singapore's post-65 military past offers an expanded universe that presents a rich environment for the scripting a 'new military history that goes beyond the themes of colonialism and 'vulnerability-survival'. Finally, just as the scripting and interpretation of Singapore's military history is an ongoing process, it is crucial that present and future generation of Singapore's military professionals pursued the life-long development of the 'historical' and 'strategic mind' in width, in depth and in context.

Endnotes

[1] Wang Gung Wu, "Contemporary and National History: A Double Challenge," in Wang Gung Wu (ed.), *Nation-Building: Five Southeast Asian Histories* (Singapore: Institute of Southeast Asian Studies, 2005), p. 5.

[2] D. R. SarDesai, *Southeast Asia: Past and Present* (Boulder, Colorado: Westview Press, 2003), p. 149.

[3] Kevin Blackburn and Karl Hack, *War Memory and the Making of Modern Malaysia and Singapore* (Singapore: NUS Press, 2012), p. 342.

[4] Karl Hack and Kevin Blackburn, *Did Singapore Have to Fall?: Churchill and the Impregnable Fortress* (London: RoutledgeCurzon, 2004), pp. 182–183.

[5] Nicholas Tarling, *Southeast Asia: A Modern History* (Melbourne: Oxford University Press, 2001), p. 290.

[6] Blackburn and Hack, *War Memory and the Making of Modern Malaysia and Singapore*, pp. 98–101.

[7] Kevin Blackburn and Chew Ju Ern, "Dalforce at the Fall of Singapore in 1942: An Overseas Chinese Heroic Legend," *Journal of Chinese Overseas*, vol. 1, no. 2, Nov 2005, p. 240.

[8] A.J. Stockwell, "Southeast Asia in War and Peace: The End of European Colonial Empires," in Nicholas Tarling (ed.), *The Cambridge History of Southeast Asia Volume Two, Part Two, From World War II to the present* (Cambridge: Cambridge University Press, 1992), pp. 1–58.

[9] Ho Shu Huang, "Rethinking the who, what and when: Why not Singaporean military heroes?" in Norman Vasu, Yolanda Chin and Kam-yee Law (eds.), *Nations, National Narratives and Communities in the Asia-Pacific* (London and New York: Routledge, 2014), p. 14.

[10] Blackburn and Hack, *War Memory and the Making of Modern Malaysia and Singapore*, p. 216.

[11] Malcolm H. Murfett, John N. Miksic, Brian P Farrell and Chiang Ming Shun, *Between Two Oceans: A Military History of Singapore from First Settlement to Final British Withdrawal* (Singapore: Marshall Cavendish Academic, 2005), p. 347.

[12] Lee Kuan Yew, *From Third World to First: The Singapore Story: 1965–2000* (Singapore: Singapore Press Holdings, 2000), p. 33.

[13] Robert M. Citino, "Military "Histories Old and New: A Reintroduction," *American Historical Review*, October 2007, p. 1071.

[14] Ramachandran Menon (ed.), *One of A Kind: Remembering SAFTI's First Batch* (Singapore: SAFTI Military Institute, 2007), p. 27.

[15] Tommy Koh and Chang Li Lin (eds.), *The Little Red Dot: Reflections by Singapore's Diplomats* (Singapore: World Scientific, 2005).

[16] *New Straits Times*, "Vital for Singapore's Leadership to address the discomfort of a Neighbour" 9 March 2002, http://www.singapore-window.org/sw02/020309n1.htm.

[17] Derek da Cunha, "Defence and Security: Evolving Threat Perceptions," in Derek da Cunha (ed.), *Singapore in the New Millennium: Challenges Facing the City-State* (Singapore: Institute of South East Asian Studies, 2000), p. 135.

[18] Tim Huxley, *Defending the Lion City: The Armed Forces of Singapore* (Sydney: Allen and Unwin, 2000), pp. 31–32.

[19] Tan Siok Sun, *Goh Keng Swee: A Portrait* (Singapore: Editions Didier Millet, 2007), p. 128.

[20] Mickey Chiang, *SAF and 30 Years of National Service* (Singapore: Ministry of Defence, 1997), pp. 57–61.

[21] Huxley, *Defending the Lion City*, pp. 56–58.

[22] Bernard Loo, "Goh Keng Swee and the Emergence of a Modern SAF: The Rearing of a Poisonous Shrimp," in Emrys Chew and Kwa Chong Guan (eds.), *Goh Keng Swee: A Legacy of Public Service* (Singapore: World Scientific Publishing, 2012), p. 140.

[23] Huxley, *Defending the Lion City*, p. 58.

[24] Yee-Kuang Heng and Ong Weichong, "The Quest for Relevance in Times of Peace: Operations Other Than War and the Third-Generation Singapore Armed Forces," in Chiyuki Aoi and Yee-Kuang Heng (eds.), *Asia-Pacific Nations in International Peace Support and Stability Missions* (New York: Palgrave Macmillan, 2014), p. 146.

[25] Kumar Ramakrishna, *"Original Sin"? Revising the Revisionist Critique of the 1963 Operation Coldstore in Singapore* (Singapore: Institute of Southeast Asian Studies, 2015), p. 126.

[26] Karl Hack and Jean-Louis Margolin, "Singapore: Reinventing the Global City," in Karl Hack and Jean-Louis Margolin with Karine Delaye (eds.), *Singapore from Temasek to the 21st Century: Reinventing the Global City* (Singapore: NUS Press, 2010), p. 4.

[27] Greg Kennedy, "Symbol of Imperial Defence: The Role of Singapore in British and American Far Eastern Strategic Relations, 1933–1941," in Brian Farrell and Sandy Hunter (eds.), *Sixty Years On: The Fall of Singapore Revisited* (Singapore: Eastern Universities Press, 2002), p. 42.

[28] *Defending Singapore in the 21st Century* (Singapore: Ministry of Defence, 2000), pp. 6–7.

[29] Geoffrey Till, *Seapower: A Guide for the Twenty-First Century* (London and New York: Routledge, 2009), p. 9.

[30] John N. Miksic, "Singapore's Maritime Heritage," in *A Maritime Force for a Maritime Nation: Celebrating 50 Years of the Navy* (Singapore: Straits Times Press, 2017), p. 18.

[31] Malcolm H. Murfett, John N. Miksic, Brian P. Farrell and Chiang Ming Shun, *Between Two Oceans: A Military History of Singapore from 1275 to 1971* 2nd Edition (Singapore: Marshall Cavendish Editions, 2011), pp. 1–35.

[32] *Ibid.*

[33] RSN50: Making Waves, Cyberpioneer, 1 May 2017 https://www.mindef.gov.sg/web/portal/pioneer/article/cover-article-detail/ops-and-training/2018-q1/rsn50-making-waves.

[34] "Two Thousand Two Hundred and Sixty-Three Days 2007–2013" (Singapore: Ministry of Defence, 2013), p. 24.

[35] Ong Weichong, "The Expeditionary Role of the Singapore Armed Forces," *Defence Studies*, vol. 11, no. 3, September 2011, p. 548.

[36] *2263 Days Operation Blue Ridge The SAF's Six-year Mission in Afghanistan* (Singapore: Ministry of Defence, 2013).

[37] The Humanitarian Assistance Support Group (HASG) to Meulaboh, Sumatra, Indonesia included soldiers from the 7th Singapore Infantry Brigade (7 SIB), the SAF Medical Corps, combat engineers, drivers, commandos, signallers and logisticians. The RSN deployed three LSTs—RSS Endurance, RSS Endeavour and RSS Persistence—along with other support vessels and naval divers. The RSAF's deployed C-130 and Fokker-50 fixed-wing aircraft and Chinook and Super Puma helicopters. Derek Liew, "2004—Operation Flying Eagle," *This Month in History*, vol. 9, no. 12, December 2005.

[38] 7th Singapore Infantry Brigade (7 SIB) consisting of a heliborne battalion and another amphibious battalion from the Guards Formation is an active service component of 21st Division, Singapore's main Rapid Deployment Force (RDF). Other components of 21st Division include a reservist heliborne brigade, a reservist amphibious brigade, an active amphibious mechanized battalion and heli-portable divisional artillery.

[39] David Boey, *Reaching Out: Operation Flying Eagle* (Singapore: SNP International, 2005).

[40] Gwee Li Sui, "NS in Literature: The Write of Passage," in *Giving Strength to our Nation: The SAF and Its People* (Singapore: Ministry of Defence, 2015), p. 363.

[41] Seah Tzi Yan, "SAF in Art: On a Broader Canvas," in *Giving Strength to our Nation: The SAF and Its People* (Singapore: Ministry of Defence, 2015), pp. 365–368.

[42] Michael Howard, *The Causes of War* (London: Unwin Paperbacks, 1983), p. 217.

[43] *Ibid.*, p. 209.

[44] Eliot Cohen, "The Historical Mind and Military Strategy," *Orbis*, vol. 49, no. 4, Fall 2005, p. 575.

[45] Richard Hart Sinnreich, "Awkward partners: Military history and American military education," in Williamson Murray and Richard Hart Sinnreich (eds.), *The Past as Prologue: The Importance of History to the Military Professional* (Cambridge University Press, 2006), p. 70.

[46] Cohen, "The Historical Mind and Military Strategy," p. 579.

[47] Hal Brands and William Inboden, "Wisdom without tears: Statecraft and the uses of history," *Journal of Strategic Studies*, vol. 41, no. 7, 2018, p. 25.

[48] Carl von Clausewitz, *On War* (Michael Howard and Peter Paret, trans. and ed.) (Princeton: Princeton University Press, 1984), pp. 170–174.

Chapter 4

Military Modernization in the 21st Century Problems and Prospects for Small Military Organizations

Bernard F.W. Loo

Senior Fellow, Institute of Defence and Strategic Studies,
S Rajaratnam School of Internal Studies,
Nanyang Technological University, Singapore

isfwloo@ntu.edu.sg

How do small military organizations engage in the process of military modernization? The intuitive answer, surely, is that the process of military modernization is the same, regardless of the size of the military organization in question. The process, surely, will reflect the concept of the arms dynamic, which is "the whole set of pressures that make states both acquire armed forces and change the quantity and quality of the armed forces they already possess."[1] Buzan believes the concept is universally applicable: "The term is used not only to refer to a general global process, but also to enquire into the circumstances of particular states or sets of states."[2]

However, assuming that the concept is universally applicable is one thing, understanding the specific pressures that individual states face in changing the quantity and quality of the military forces they possess is a separate question, and one that deserves an answer. This chapter proposes that the military modernization process in any military organization ought to be underpinned by three sets of questions, the answers to which would surely shape the specific modernization of the military organization. The first set of questions revolves around the geopolitical environment in which the particular state exists, the types of security challenges and threats that the state is likely to face, and the types of capabilities that the military organization needs, as a consequence. The second set of questions revolves around technological issues: the capacity of the military organization in question to keep pace with, as well as absorb, new and increasingly exotic military and military-related technologies; and the extent to which the existing capabilities of the military organization create a technological path dependency, locking the military organization into either the United States-centric or Russia-centric technological universes. The final set questions revolve around the issue of affordability, and the attendant choices thereafter: what is the relative and projected health of the economy, and what percentage of government spending can be realistically dedicated to military spending, and specifically to the modernization of the military organization?

Through a case study of the military organizations of the ten states of Southeast Asia, this chapter examines the pressures that these military organizations will face in the not-too-distant future. It makes three arguments. One, it argues that technological change—and sometimes, radical change—is unavoidable for most, if not all, military organizations. The challenge will be to adapt what is still a predominantly foreign phenomenon to their local strategic circumstances. Two, attempting to keep apace of the rate of technological change will be a very difficult process for these military organizations. Military technologies are increasingly exotic, and will likely move military organizations into uncharted and potentially dangerous waters. For small military organizations, a wrong turn may be potentially catastrophic. Furthermore, the issue of affordability will become increasingly important, given the escalating costs of new technologies and weapons systems.

The third argument this study makes is that acquiring the latest military technologies may result in military organizations that are at odds with the dominant strategic challenges in the region for the foreseeable future. As this chapter will later demonstrate, there is a growing body of opinion that the security agenda for many states—and in this chapter's case, Southeast Asian states—has for some time been moving away from traditional state-oriented concerns about war towards so-called non-traditional security concerns such as peace operations, low-intensity counter-insurgency and counter-terrorism operations, and humanitarian and disaster relief operations stemming from pandemics and natural disasters. Before this argument can be constructed, however, it is necessary to examine Buzan's concept of the arms dynamic, and discuss the extent to which it helps us to understand specific military modernization program in individual countries.

The Concept of the Arms Dynamic

The concept that encapsulates the various military modernization processes of military organizations around the world is the arms dynamic. Barry Buzan produced the first full-length discussion of the concept in his 1987 book, *An Introduction to Strategic Studies*. Together with Eric Herring, Buzan would refine the concept in *The Arms Dynamic and World Politics*, published in 1998. If the arms dynamic is "the whole set of pressures that make states both acquire armed forces and change the quantity and quality of the armed forces they already possess," it is important to identify the pressures—and these pressures are exogenous and endogenous in nature—that drive the arms dynamic.

The exogenous drivers of the arms dynamic are the action-reaction model. "States will arm themselves either to seek security against the threats posed by others or increase their power to achieve political objectives against the interest of others."[3] Because of the tendency to equate the action-reaction model with the concept of arms races, it can be tempting to associate the action-reaction model exclusively with inter-state relations that are inherently and overtly hostile. However, Buzan and Herring argue that the model still "works at the lower levels of competition" because state actors "usually have some sense of who they consider to be possible sources of attack even when they see the probability of war as being low."[4] Finally, despite its association with arms racing, the action-reaction model does not preclude the possibility of arms reductions: in a context where a state's military planners and policy makers conclude that the probability of the state being embroiled in armed conflict is so low as to be negligible, the logical conclusion is that those policy makers can decide to reduce military expenditures and reduce the size and capability of the state's military organization.

The endogenous drivers of the arms dynamic are the domestic structures model: where "the process of the arms dynamic has become so deeply institutionalized within each state that domestic factors largely supplanted the crude forms of action and reaction as the main engine of the arms dynamic."[5] The domestic structures model does not necessarily preclude the action-reaction model, although it is hypothetically possible for a specific case where "the military behavior of states is generated much more by internal considerations than by any rational response to external threats."[6] More likely, there is a balance between the two models that nevertheless tilts in favor of endogenous drivers and/or considerations, such as: the institutionalization of military expenditure, research and development, and production and acquisition; organizational politics within the military; or electoral politics within the state.

Where the 1987 and 1998 books diverge is in the importance placed on the influence of technology. In his 1987 book, Buzan had seen technology—more precisely, the "qualitative evolution of technology as a whole"—as a "fundamentally independent element of the arms dynamic that is not fully captured by either action-reaction or domestic structures."[7] By 1998, Buzan and Herring removed the technological imperative as a separate set of pressures on the arms dynamic, and instead came to see a technological imperative as permeating both action-reaction and domestic structures models: "Although action-reaction and domestic structure factors do play a substantial role in [the arms dynamic], it is important not to lose sight of the point that both sets of factors are themselves heavily conditioned by the independent process of the technological imperative."[8]

The Centrality of Technological Change

Technology—and specifically, technological change—is integral to the arms dynamic, as noted earlier. The military historian, Martin van Creveld, has argued that technology exercises a pervasive influence over strategy and war, and this means that as military technologies change, so too must the machinery of war undergo periodic change.[9] Military capabilities change over time, and on occasions, it might be necessary to review the modus operandi of the military organization, especially if new types of capabilities avail themselves and these new capabilities require radically different modus operandi for their effects to be maximized. In the 21st Century, one set of technologies arguably underpins most if not all of military technologies, namely, information technologies in the form of computing and networked digital communications. It is worth noting here

Table 1:   Major Suppliers, Global Arms Market, 1990–2019.

Rank	Supplier	1990	2000	2010	2019	1990–2019
1	USA	10733	7577	8033	10752	289310
2	Russia	—	4503	6275	4718	145717
3	Germany	1856	1612	2664	1185	53071
4	France	1818	1082	866	3368	52926
5	UK	1866	1633	1157	972	38878
6	PRC	941	299	1478	1434	29777
7	USSR	9895	—	—	—	15535
8	Netherlands	416	284	371	285	15431
9	Italy	206	208	539	491	14865
10	Israel	85	404	637	369	14171
	Top 50	30175	19257	25732	27170	759632
	Others	198	83	40	28	2715
Total		**30370**	**19337**	**25771**	**27194**	**762349**

Source: Stockholm International Peace Research Institute, Arms Transfers Database.

that these technologies are ubiquitous, and available to any actor who has the ability to purchase them. Furthermore, in the transparent environment of a globalized Internet world, scientific breakthroughs are shared rather quickly and there are few sustained exclusive monopolies on technology. There are three issues that small military organizations have to address in facing up to the issue of technological change.

One, the global arms market is overwhelmingly dominated by a small number of suppliers of military technologies. Table 1 lists the top ten countries in the global arms market in the period of 1990 to 2019. The United States is the single largest player in the global arms market, its market share amounting to almost 40%. Russia, the country with the second largest share of the market, accounts for less than 20% of the market. According to the Stockholm International Peace Research Institute, only two Southeast Asian countries—Indonesia and Singapore—are within the ranks of the top 50 exporters of military equipment.

Two, it is important to remember that all military technologies come from specific, even unique, geopolitical and geostrategic circumstances. The technologies that allowed the United States and its allies to defeat Iraqi military forces in Operation Desert Storm—often understood under the concept of the Revolution in Military Affairs (RMA)—emerged in the 1970s, during the height of the Cold War, in response to the central strategic problem facing the United States and its North Atlantic Treaty Organisation (NATO) partners: that its adversary, the Warsaw Pact, enjoyed significant numerical advantages against NATO in heavy armored forces. Then-Secretary of Defense, Harold Brown, envisioned the use of technological force multipliers—information technology-enabled capabilities such as pervasive precision sensing and precision targeting, and battlefield management systems underpinned by digital networked communications—to nullify the Warsaw Pact's significant numerical advantages.[10] Currently, the military capabilities that these major exporters produce, much of which is exported to other countries, have been designed to meet the strategic challenges that they currently face. The export of these military capabilities is as much about geopolitical considerations (exporting to alliance partners and other like-minded or friendly countries) as it is about economic considerations (generating economies of scale for efficient production, trade balances). Furthermore, there is a technological divide between the United States and Europe on the one hand, and Russia and the People's Republic of China on the other, and the two technological standards do not integrate with each other. This can result in path dependencies: a country that has traditionally bought its military equipment from the former Soviet Union or the People's Republic of China may find it difficult to acquire military equipment from the United States, at least if it intends to integrate the different technological standards into a coherent and functioning military system.

Three, affording these military technologies may impose every-increasing demands on national economies. There appears to be a general sense that the costs of weapons systems have been more or less trending upwards. David Kirkpatrick consistently maintained that weapons systems costs have been, and will continue to, increase.[11] A 2007 study by RAND Corporation found the development costs of weapons programs since the 1990s has been increasing.[12] Current and emerging weapons systems, especially air and naval weapons systems, appear to be expensive. The F-35 Joint Strike Fighter program is widely regarded as one of the most expensive weapons acquisition programs ever; the Lot 14 batch of F-35s across the A to C variants cost US$78 million, US$101 million, and US$94 million respectively.[13] The question is whether or not it is a one-off phenomenon. Russian weapons systems are apparently cheaper: in June 2020, India reportedly concluded an agreement to acquire 21

MiG-29 and 12 Su-30 combat aircraft for a total of US$2.4 billion. Nevertheless, even current Russian weapons systems are more expensive than their predecessors: the Armata T-14 main battle tank costs have been such that the Russian defence ministry decided against using it as a replacement for its legacy main battle tanks currently in service.[14] Indeed, rising costs of new weapons systems, against of backdrop of limited economic growth, might even result in structural disarmament: a hypothetical scenario is where 50 obsolete combat aircraft are replaced by 20 new combat aircraft.[15]

These three issues can be seen in Southeast Asian arms dynamics. Table 2 outlines the countries of Southeast Asia, and the major suppliers of land, naval and air combat weapons systems. Three observations about Southeast Asia's arms dynamics are germane.

One, there are the remnants of the Cold War divide in Southeast Asia. The Indochinese states (Cambodia, Laos, Myanmar and Vietnam) have tended to opt for the Russian technological universe (including weapons systems from the People's Republic China and other Warsaw Pact states). On the other hand, the military organizations of maritime Southeast Asia (Brunei, Indonesia, Malaysia, the Philippines, Singapore and Thailand) have traditionally relied on either the United States or other NATO countries for much of their military hardware. This is especially true with regards to high-value, big-ticket items such as advanced air combat or

Table 2:   Major Suppliers, Southeast Asia Market, 1990–2019.

Country	Land	Naval	Air
Brunei	France	Germany	USA
Cambodia	France	PRC	PRC
Indonesia	France	Germany/Netherlands	Russia/USA
Laos	Russia	—	PRC/Russia
Malaysia	—	France/Germany	Russia/USA
Myanmar	PRC	PRC	PRC/Russia
Philippines	USA	USA	USA
Singapore	Germany	France/Germany	USA
Thailand	USA	PRC/USA	USA
Vietnam	Russia	Russia	Russia

Source: Stockholm International Peace research Institute, Arms Transfers Database.

naval platforms. Recently, Malaysia and Indonesia have begun to look towards Russia as a source of high-technology, high-value items such as combat aircraft, and this trend looks likely to continue.[16] That being said, neither country appears to have decided to migrate totally into the Russian technological universe—the recent acquisitions in naval combat platforms as an example.

Two, the dominance of the United States and Russia in terms of sales of advanced combat aircraft also means that the countries of Southeast Asia are hostage to the geostrategic concerns that these major players in the global arms market have in designing their respective combat aircraft. The F-15s and F-16s, MiG-29s, and Su-27s, Su-30s and Su-35s that a number of Southeast Asian air forces currently operate were designed to address the geostrategic interests and concerns of the United States and Russia respectively. Whether or not these advanced combat aircraft are geostrategically for the countries of Southeast Asia is potentially open to debate. Furthermore, the emerging defense technologies in these major suppliers are moving into increasingly exotic technological domains: directed energy weapons, hypersonic missiles, artificial intelligence-enabled weapons systems.[17] Integrating such exotic technologies into military organizations that have barely come to grips with what Stephen Biddle calls the modern system[18] may be an insurmountable challenge for those Southeast Asian military organizations that still rely on weapons systems bordering on decrepitude. The Philippines, for instance, has traditionally relied on the United States for its military equipment; whether United States military technologies today are relevant to the strategic challenges facing the Philippines remains to be seen. Can these Southeast Asian military organizations break free of their historical dependencies in the area of military modernization? If they cannot, can they remain relevant to the strategic challenges facing their respective states. This question will be examined in greater detail later in this chapter.

Three, while Southeast Asian defense spending has generally increased between 1990 and 2019, this increase is reflected solely in absolute amounts; when expressed as a percentage of the countries' respective gross domestic products, however, a different picture begins to emerge. The figures provided within parentheses in Table 3 indicate the countries' defense spending as a percentage of GDP; the striking feature in this regard is the relative stability, and even stagnation, of military spending, in terms of individual countries' national priorities. This stability presumably reflects Southeast Asian policy makers' perceptions of a relatively

Table 3:   Military Expenditures, Southeast Asia, 1990–2019.

	1990		2000		2010		2019	
Brunei	395	(6.4)	326	(3.90)	392	(2.9)	419	(3.3)
Cambodia	12	(2.1)	166	(1.5)	211	(1.5)	593	(2.3)
Indonesia	2423	(1.4)	2228	(0.7)	4369	(0.6)	7380	(0.7)
Laos	—	—	35	(0.6)	20	(0.2)	—	—
Malaysia	1602	(2.6)	2164	(2.2)	3713	(1.5)	3827	(1.0)
Myanmar	511	(3.4)	653	(1.1)	—	—	—	—
Philippines	1943	(2.1)	2173	(1.5)	2641	(1.2)	3327	(1.0)
Singapore	3835	(4.6)	7391	(3.9)	9325	(3.4)	11262	(3.2)
Thailand	3963	(2.6)	3395	(1.4)	5473	(1.6)	6970	(1.3)
Vietnam	2009	(7.9)	—	—	3485	(2.3)	—	—

Source: Stockholm International Peace Research Institute, Military Expenditure Database.

stable and benign regional security environment; presumably this means that if these perceptions of the security environment change for the worse, that defense spending can begin to command a greater share of the national "economic pie." Nevertheless, all things being equal, if defense spending does not increase in relative terms, in a context where costs of new military technologies are increasing, the phenomenon of structural disarmament, mentioned earlier, may emerge in Southeast Asia. Indeed, it is plausible that structural disarmament has already occurred for at least one country in Southeast Asia: Singapore has decided to acquire four F-35Bs, with an option to acquire a further eight aircraft, to replace the F-16s currently in operation with the Singapore Air Force. It is not entirely clear that the 60 F-16s in the Singapore Air Force inventory will be replaced by a similar number of F-35s; and given the F-35B's price tag, it seems more likely that the eventual number of F-35Bs in the Singapore Air Force will be significantly less than the 60 F-16s.

Against this pattern of ever-increasing costs of new weapons systems and platforms, even the advanced countries find the required expenditures prohibitive. The identified solution has been to move towards off-the-shelf purchase (often accompanied by licensed production) and collaborative procurement or weapons research development (such as the Eurofighter Typhoon and the F-35 Joint Strike Fighter). Such international collaborations appear to be an attractive option, but this can be problematic.[19] These measures are designed to mitigate the problem of affordability of new military technologies. Nevertheless, the point remains: for the military organizations of Southeast Asia, engaging with current military technologies will require a significant injection of financial resources, which may be something that most Southeast Asian states will be unable to afford.

Matching New Technologies with New Techniques

Being able to keep pace with technological change, and being able to afford new technologies, is one set of issues. Another set of issues, the focus of this section, is the ability to properly integrate new technologies within the military organization to achieve tactical and strategic effectiveness.

The rate of technological change in the non-military domain can be breath-taking. The Voyager spacecraft have under 70 KB of memory, in contrast to the first iPhone, which had a minimum of 4 GB of memory.[20] Furthermore, the first iPhone (released in 2007) had 128 MB of RAM, a CPU speed of 400 MHz; iPhone 11 (released in 2019) has 4 GB of RAM, and a processing speed of up to 2.65GHz. Outside of the home, the cutting edge of information technology today revolves around artificial intelligence and quantum computing, the latter technology having emerged in the second decade of the 21st Century, and already outperforming the best supercomputers that had become a stable technology in the first decade of the 21st Century. In the military domain, technologies that were considered "exotic" two decades ago—Blue Force Tracking, GPS navigation, night vision devices—are now standard-issue for many military organizations. The pace of change in computing technologies is also being felt in the military domain. What new technologies will emerge from over the horizon—and in particular, when and how fast these new technologies will evolve—cannot be accurately determined. But extrapolating from Moore's Law to the rate of overall technological change, it is almost inevitable that the pace of technological change will only accelerate. The rates at which new technologies are introduced to military organizations will almost certainly also accelerate. What the pace of technological change can mean for military organizations is the potential acceleration of technological obsolescence of extant military capabilities.

The technological path is one issue; there are other paths that need to be negotiated. A key argument of the RMA literature of the late 1990s onwards was that the technological changes alone were not sufficient to constitute a true "revolution"; rather operational concepts and doctrines, organizational structures and cultures, these would all have to undergo radical, disruptive change as well.[21] While the RMA concept might now be passe, a brief discussion of its recommendations in the conceptual and doctrinal, as well as organizational, areas is illustrative of the issues facing defense planners today as they address military modernization in the 21ˢᵗ Century.

This largely American vision encompasses four main pillars: enhancing joint operations, exploiting intelligence advantages, experimenting with new concepts, and developing transformational capabilities.[22] At the organizational level, much was also made about the need to abandon long-standing single service cultural traditions in favor of networked joint forces. Furthermore, networked digital communications would encourage a flatter hierarchical structure. Layers of middle management may need to be removed. On a related note, there were calls for radical organizational re-structuring, moving the military organization from its traditional, industrial-era hierarchical and functional structures towards flatter, less hierarchical and less stove-piped structures. These technologies therefore challenge the continued primacy of existing (and often stubborn) hierarchical structures and command cultures in military organizations. If the functional structures have to be revamped, this will have implications for the nature of command. But this is only one possible outcome; connectivity works both ways, and increased vertical connectivity does not merely result in better upwards information flows, it also facilitates greater top-down micromanagement.[23] As a result of greater connectivity and flatter hierarchies, the distance between upper and lower echelons is reduced, this may facilitate the empowerment of lower echelons to independent actions all within the aim and intent of the operation (so-called *Auftragstaktik*); but the shorter distance between supreme command and foot soldier also means it is increasingly possible for higher command to intervene in a single foot soldier's actions.

In such an organizational environment, extant operational concepts and doctrines would have to change. Old ways of strategic thinking had to be replaced. The RMA forwarded "new" concepts like network-centric operations and effects-based operations and full-spectrum dominance, to challenge the extant doctrinal and tactical orthodoxy. *Joint Vision 2020* offered the vision of United States military forces "operating unilaterally or in combination with multinational and interagency partners,

[able] to *defeat any adversary* and *control any situation* across the full range of military operations."[24] The document further states that "the joint force of 2020 will seek to create a "frictional imbalance" in its favor by using the capabilities envisioned … but the fundamental sources of friction cannot be eliminated."[25] Additionally this joint force of 2020 "will use superior information and knowledge to achieve decision superiority, to support advanced command and control capabilities, and to reach the full potential of dominant maneuver, precision engagement, full dimensional protection, and focused logistics."[26]

Nevertheless, there have been concerns about such radical departures, whether in terms of organizational structures or in doctrines and operating concepts. The RMA promised a military organization that is an increasingly lethal and capable of unprecedented levels of precise destruction.[27] There is an argument to be made that lethality and destructive capacity are elements of a uniquely American phenomenon, one that fits naturally with American military culture and the American way of war fighting. Antulio Echevarria depicts the American way of war as one that "tends to shy away from thinking about the complicated process of turning military triumphs, whether on the scale of major campaigns or small-unit actions, into strategic successes."[28] American military culture celebrates campaign and battle victories, and has resulted in a unique and specific manner of war fighting. In a similar vein, Arthur Cebrowski and Thomas Barnett highlight speed and precision as characteristics of the American military.[29] It might be possible to see the American way of war, inasmuch as it exists, as part of a broader western way of war.[30] However, the American military's lethality and destructive power has not always translated into strategic effectiveness, which is the ability to impose one's political objectives on an opponent. Indeed, there is a case to be made that the United States military may have secured spectacular battlefield victories in its military operations in Afghanistan and Iraq, but in both theaters, have nevertheless failed to secure American political interests in these countries.[31] Furthermore, such radical departures are, for a variety of reasons, difficult to accept, not least because radical departures require great leaps of faith into the unknown. A study by Thomas Mahnken and James FitzSimonds highlighted the mixed reactions within the United States land forces to the notion of military transformation.[32] If the RMA leads down the wrong path, then for small military organizations, the potential consequences of going down such a wrong path may be devastating.

Where do the military organizations stand in terms of their individual capacities to absorb new technologies, and possibly initiate radical changes in organization and/or operational concepts? It can be argued that the proposition of radical organizational change comes into potential conflict with the strategic cultural and geopolitical conditions that Southeast Asian military organizations find themselves in. At one level, the military organizations of Southeast Asia may more or less resemble their European counterparts, albeit superficially, in terms of their internal organization and nominal responsibilities.[33] At another level, however, the roles of Southeast Asia's military organizations exceed those of conventional Western military organizations. Many military organizations in Southeast Asia region have tended to be associated with internal security. The need to maintain internal security in the face of such challenges to state authority was often the primary *raison d'etre* for Southeast Asian military organizations. This historical legacy formed the backdrop to the development and expansion of the regional military organizations up till the end of the Cold War.[34] Clearly, the dominant American vision of the transformed military organization has the potential to run around against this vision of the Southeast Asian military organization.

Secondly, the fact that the information technologies discussed earlier are ubiquitous should not rule of the question of whether or not the military organizations of Southeast Asia can, or ought to, engage with these technologies. The specific geopolitical origin of current exotic military technologies means that it is perfectly reasonable to question its universality today. In the case of the RMA, the Australian strategist, Paul Dibb, argued that there were three key discriminators: the relationship to the United States, the capacity to absorb these technologies, and threat perception.[35] His conclusion was that some states, believing themselves to exist in high-threat environments, may decide that the RMA is necessary, but this will not necessarily be the case for other states who perceive themselves to exist in relatively more benign security environments. Simply put, not every state requires military organizations equipped with cutting-edge military technologies.

Dibb's analysis was centered on the RMA and Asia Pacific region, but the analysis is equally applicable to any region and to current post-RMA military technologies. In the 21ˢᵗ Century, any military modernization program will require a number of skillsets of the pertinent organization. The first is systems integration skills, which are the most demanding aspect, since they place great demands on the country's education system to nurture those skills and the qualities of creativity, innovation and independence of thinking. Secondly, the development of new operating concepts and doctrines will almost certainly have to be out in place. Thirdly, joint force operations require an integrated logistic support and maintenance structure. Finally, the systems integration challenge is exacerbated by the proliferation of different technologies and technological standards, a practice that is quite common in smaller military organizations as they source for military hardware that they can afford. These challenges come crashing together in the case of many military organizations in Southeast Asia. Political cultures tend to favor conformity and deference to authority. Weapons systems are sometimes acquired without due consideration to developing the doctrines needed to seamlessly integrate these weapons systems into existing or new force structures. Separate single-service cultures tend to be the norm in Southeast Asia. Several military organizations in the region also have weapons systems acquired from both United States/NATO as well as Russian sources. Integrating disparate weapons systems may prove to be an insurmountable challenge. If nothing else, this presents such military organizations with tremendous logistical headaches.

Maintaining the Relevance of the Military Organization

Finally, it is important to ask if the modernized military organization is necessarily relevant to the strategic challenges that the state faces. By the end of the 20th Century, a debate emerged as to the nature of the security agenda in the 21st Century for states and military organizations.[36] One side argues for a new security agenda, pointing to phenomena such as "new" wars, fourth generation (possibly even fifth) warfare and cyber warfare on the one hand. The other side holds to a "business as usual" argument: that the grammar of war (its specific manifestations) may change, but the logic of war (its essence) will remain unchanged. However, if we assume that the new security agenda thesis to be true, this then begs the question of the extent to which military organizations will be prepared for such scenarios. Some discussion of this proposition is warranted.

This new security agenda thesis portrays the changes to international politics as a paradigm shift, a radical reorientation of how we perceive and think of the world around us. One particular aspect of this reorientation of international politics, germane to the analysis in this chapter, is the expansion in the number and power/influence of malign non-state actors, in particular: transnational criminal organizations; and rebel groups and terrorist organizations and networks. Transnational criminal organizations have been waging what Moises Naim, the former editor of *Foreign Policy*, called the five "wars" of globalization—the proliferation of narcotics, human trafficking, intellectual property theft, money laundering, and arms smuggling and black market trade.[37] In the case of rebel groups, their strategic significance is reflected in statistics that demonstrate that major inter-state wars (however "major" is defined) has been on the wane since World War II, whereas so-called "small" inter-state and intra-state wars have been on the rise.[38] Finally, while some studies suggest that mainstream media has overhyped the threat of terrorism to state and human security, other studies suggest that between 2002 and 2014, terrorist attacks increased significantly.[39] Then actors are considerably more agile than the cumbersome, slow state bureaucracies they face. Today's terrorists and insurgents are as, if not more, computer-savvy as their state adversaries. Information technology is a double-edged sword. This is an era in which the bulk of military capabilities are increasingly civilian developed, relatively cheap, commercially available and globally distributed. Terrorists and insurgents are also able to access technology that is commercially available to carry out operations against states that have strong military forces.[40] An increasing number of individuals and non-state actors as well as states can get access to overhead imagery, night vision devices, biological weaponry, thermal image defeating materials, robotic vehicles (land, sea and air), system integration software, micro-satellites, sophisticated communications and conventional weaponry of all kinds. The means to wage war are no longer the exclusive domain of states.

Given this backdrop, it is fair to question the continuing relevance of military organizations, especially those that are configured principally for conventional military operations. Military organizations have traditionally been configured to deal with the traditional security challenges, centred on the threat of invasion by other states, concerns of territorial integrity and sovereignty. Deploying the military organization *du jour* for these new security concerns may be necessary, but this prospect may also be inherently problematic.[41] Furthermore, these new security concerns require skill sets different from those that conventional military operations demand, which typically do not occupy very much attention in the training regimes of modern militaries.[42]

However, a contrary view, which rejects the notion that conventionally-structured military organizations lack the skill sets to address the non-war security challenges of the 21st Century earlier posited, can be held. Where terrorist and insurgent facilities can be located, these can be attacked and destroyed by increasingly very precise instruments of military power.[43] Even the more passive counter-terror and counter-insurgency measures—such as the guarding of critical infrastructure—resonate with military organizations around the world. Nevertheless, a caveat needs to be introduced. While traditional military skill sets are applicable to counter-terrorism operations, great care must be taken to tailor these traditional skill sets to the unique conditions of the particular counter-terrorism operation. The application of military force against terrorist bases has to be very carefully calibrated, so as to not incur unnecessary levels of destruction, especially collateral damage. This, of course, is what current military technologies—with their emphasis on precision targeting—excels in. Nevertheless, restraint in the use of force is desirable,[44] but this may run against the grain of the military mindset.

Finally, another increasing operational demand on military organizations is humanitarian and disaster relief operations. In a way, this is not new, as it has been fairly commonplace for governments to mobilize its military organization in times of natural disaster. Military organizations have skill sets and capabilities that are relevant to such operations: the organization and management of logistics, and the capacity to move men and materiel over long distances. The United States military's deployment in the wake of Hurricane Katrina is merely a more recent manifestation of this requirement. The Chinese government has regularly used the People's Liberation Army in times of major natural disasters. What is fairly new, however, is the increasing willingness of countries to deploy their respective military organizations to other countries who have suffered natural disasters. Witness, for instance, the global relief efforts that came in the wake of recent disasters: the December 2004 tsunami that affected parts of Indonesia, Bangladesh, Sri Lanka, Thailand and India; the 2008 Sichuan earthquake, and the willingness of countries such as Japan, Russia and Singapore to contribute manpower and materiel to support Chinese rescue efforts; and the 2008 Cyclone Nargis disaster that hit Myanmar and Sri Lanka, and the efforts of Malaysia and Thailand, amongst others, to provide for disaster relief. Nevertheless, can military modernization today result in a military organization that is capable of mounting such operations as and when needed? And do these operations result in military organizations having to prioritize time away from their primary mission of protecting the state against armed aggression?

The military organizations of Southeast Asia are not immune to the potential challenge of remaining relevant to the specific security and operational demands facing military organizations identified above. If anything, it can be argued that counter-insurgency/counter-terrorism and disaster assistance and relief operations—the latter, both at home and abroad—are exactly the types of operations that Southeast Asia's military organizations have been conducting, if not exclusively, then definitely since the end of the Cold War.

The majority of Southeast Asian military organizations have had a long tradition of counter-insurgency operations. Up to the 1990s, most of the military organizations in the region were in fact configured primarily (if not exclusively) for counter-insurgency operations. Once the Cold War ended, however, and the threat of Communist insurgencies in Southeast Asia receded, many began to reconfigure themselves for conventional military operations instead. The Malaysian Armed Forces stands out as the paramount example of this type of re-structured military organization. Counter-terrorism has also become the dominant security concern for many Southeast Asian countries. In the process, however, Southeast Asian countries have to now balance their finite resources between wanting to maintain their conventional military postures and moving back to their counter-insurgency traditions. As the earlier analysis has demonstrated, military organizations such as the Malaysian Armed Forces face a number of contending push and pull factors – the trend of military technological developments, the costs of new military technologies, and the security missions that they are likely to face.

The increasing salience of humanitarian and disaster relief operations can exercise an impact on force structure planning in some Southeast Asian countries. A former Indonesian defence minister, Juwono Sudarsono, had in 2008 publicly argued for transport platforms to be the emphasis of Indonesia's military modernization programs, given the natural disasters that frequently occur in the country.[45] Even Singapore, which does not bear the brunt of the natural disasters of the region, has sought to develop a flexibility in its military organization, by acquiring a fleet of locally-designed and built Endurance-class vessels that can be used to project military forces for both humanitarian and combat operations. Such an option might represent a potentially successful attempt at balancing between these conflicting interests and tendencies. In contrast, Malaysia has maintained programs to acquire weapons platforms such as the Scorpene-class submarines, as well as expressing interest in further acquisitions in air combat platforms.[46] At the same time, however, in tune with the new security agenda argument here, Malaysia recently unveiled new maritime surveillance systems to maintain security in the Malacca Straits.[47] It appears that, at least for the moment, dual-use capabilities will dominate force structure decision-making, at least for some Southeast Asian armed forces. Nevertheless, the point that this chapter has made is: if military modernization programs are not focused on capabilities that are sufficiently flexible to be deployed for either combat or non-combat operations, these military organizations will run the risk of becoming strategically irrelevant.

Conclusion

If the military organizations of Southeast Asia are a microcosm of military organizations in small states, the main conclusion that can be drawn is that for these military organizations, the process of military modernization will be increasingly difficult, for three reasons. One, if one accepts the premise that the dominant countries in the global arms market are where the cutting edge of military technologies, it means that global military technology trends promise to move in increasingly exotic, even revolutionary, directions. Military organizations may eventually look fundamentally different in all respects from its 20th Century predecessors. Furthermore, if the pace of technological change accelerates, it means the pace of technological obsolescence similarly quickens.

Two, the dominance of these countries in the global arms market means that small military organizations may have to replace obsolete capabilities with new technologies that they will find difficult to operate and integrate into their organizations. What these military organizations are going to look like, what their force structures are going to become, the operating concepts and doctrines that they will need to attain strategic effectiveness, these are questions that remain to be answered. Furthermore, structural disarmament may be unavoidable, especially if costs of new weapons systems continue to increase at a pace that outstrips individual countries' economic growth rates. If structural disarmament occurs in large military organizations—that, presumably, have more resources at their disposal—structural disarmament can be potentially catastrophic for smaller military organizations.

Three, even if military organizations can afford these new weapons systems and can integrate these new capabilities seamlessly, the transformed military organization may be at odds with the likely security challenges that their respective countries may face in the next two decades. As much as military organizations may exist to protect their respective states from external threats, the reality for many states is that their military organizations tend to be employed for operations other than war; and these operations have different requirements from that of combat operations, especially in terms of military capabilities. Yet, it will be difficult for most states to equip their military organizations for operations other than war, and ignore the capabilities required for combat operations altogether. The challenge may be to find capabilities that are sufficiently flexible to allow these military organizations to conduct both types of operations.

Does this therefore mean that for the majority of Southeast Asian military organizations, they will have no choice but to migrate from their traditional dependence on the United States and NATO states for military hardware to Russia, the People's Republic of China and other former Warsaw Pact states? This demands an extremely painful process of abandoning legacy systems from the so-called Western powers to the new systems emanating from Russia and the like. What this also hints at, which is another difficulty that the United States military-industrial complex poses to Southeast Asian military modernization ambitions, is the issue of costs, and whether or not they will be able to remain within the ambit of the US military-industrial technological universe.

What Southeast Asian military organizations increasingly face are push and pull factors that work across purposes. The United States and western Europe have historically dominated the Southeast Asian armaments market, at least for countries like Malaysia, the Philippines, Singapore and Thailand; this legacy continues to be a pull factor for these military organizations, inasmuch as these military organizations desire to remain within the technological ambit of the United States and western Europe. Furthermore, from the experiences of recent military operations in Afghanistan and Iraq, these military technologies have a successful track record; they work "as advertised." Nevertheless, the increasing costs of American and European technologies constitute considerably powerful push factors, especially for the countries with less developed economies. Furthermore, do these military organizations have a strategic requirement for military technologies that confer increasing lethality? Or, do these organizations require less advanced, more flexible capabilities that can allow them to respond to the operational challenges they face? Will "buying American" result in military organizations that are highly unsuitable for the likely security challenges of the early 21st Century?

For some Southeast Asian military organizations, the conundrum these push and pull factors create is resolvable. At one end, Singapore, clearly, has decided that it will remain within the technological ambit of the United States and western Europe, for two reasons. One, its policy makers and strategic planners perceive the country's geopolitical environment to be potentially problematic, and the country to suffer from inherent geostrategic weaknesses.[48] Two, its economy is sufficiently developed such that it can afford to remain within the technological ambit of its traditional suppliers.[49] Nevertheless, the phenomenon of structural disarmament, discussed earlier, will almost surely apply to Singapore.

For the military organizations of Malaysia and Thailand, both of which traditionally relying on the United States and western Europe, the record demonstrates an increasing willingness to acquire from so-called non-traditional suppliers, namely, Russia and the People's Republic of China. In the 1990s, Malaysia acquired 8 F/A-18s from the United States and 8 MiG-29s from Russia, a decision that was almost surely driven by Singapore's prior acquisition of F-16s (A and B variants, block 15); nevertheless, the decision to acquire technologically incompatible (and therefore non-integrated) combat aircraft was also driven in part by the problem of affordability. Affordability is almost certainly the principal consideration for Indonesia's air fleet of American and Russian combat aircraft. Thailand's navy has vessels from both the United States and the People's Republic of China; this combination is as much about affordability as it is about the geopolitical orientations of the country's policy makers. For these countries, the acquisition of technologically incompatible weapons systems means that their respective military organizations cannot function as coherent systems, which potentially means a serious reduction of potential tactical effectiveness.

To return to the concept of the arms dynamic, it ought to be clear that the domestic structures and action-reaction models can explain much about arms dynamics in Southeast Asia, but they also fail to capture the nuances and subtleties of the respective arms dynamics in the various countries. Nevertheless, as the discussion here demonstrates, neither illustrates clearly the dilemmas facing Southeast Asia's military organizations in the 21st Century. The concept certainly fails to capture the considerations—both geopolitical and geostrategic—that can shape the specific decisions a country makes in the modernization of its military organization.

Endnotes

¹Barry Buzan, *An Introduction to Strategic Studies: Military Technology and International Relations* (Houndsmill, Basingstoke: Macmillan, 1987), p. 73. Also: Barry Buzan and Eric Herring, *The Arms Dynamic in World Politics* (Boulder and London: Lynne Rienner, 1998), p. 5.
²*Ibid.*
³*Buzan, An Introduction to Strategic Studies*, p. 76; Buzan and Herring, *The Arms Dynamic in World Politics*, p. 83.
⁴*Buzan and Herring, The Arms Dynamic in World Politics*, p. 84.
⁵*Buzan, An Introduction to Strategic Studies*, p. 94; Buzan and Herring, *The Arms Dynamic in World Politics*, p. 101.
⁶Buzan and Herring, *The Arms Dynamic in World Politics*, p. 101.
⁷Buzan, *An Introduction to Strategic Studies*, p. 105.
⁸Buzan and Herring, *The Arms Dynamic in World Politics*, p. 121.
⁹See, for instance, Martin Van Creveld, *Technology and War*, second ed. (New York, USA: The Free Press, Maxwell Macmillan Canada, Inc., 1991), p. 1.
¹⁰See Elinor Sloan, *Military Transformation and Modern Warfare* (Westport: Praeger Security International, 2008), ch. 1, for a brief but very useful overview of the history of this RMA.
¹¹David L.I. Kirkpatrick, "The Rising Unit Cost of Defence Equipment—The Reasons and The Results," *Defence and Peace Economics*, vol. 6, no. 4, 1995, pp. 263–288; David L.I. Kirkpatrick, "Trends in the Costs of Weapons Systems and the Consequences," *Defence and Peace Economics*, vol. 15, no. 3, June 2004, pp. 259–273.
¹²Obaid Younossi *et al., Is Weapon System Cost Growth Increasing? A Quantitative Assessment of Completed and On-going Programs* (Santa Monica, CA: RAND Corporation, 2007).
¹³https://www.f35.com/about/cost.
¹⁴https://thediplomat.com/2018/08/russia-will-not-mass-produce-t-14-armata-main-battle-tank/.
¹⁵Ron Matthews, "The Political Economy of Defence," in Ron Matthews (ed.), *The Political Economy of Defence* (Cambridge: Cambridge University Press, 2019), pp. 5–6; Ron Matthews and Curie Maharani, "The Defense Iron Triangle Revisited," in Richard A. Bitzinger (ed.), *The Modern Defense Industry: Political, Economic and Technological Issues* (Santa Barbara, Denver, and Oxford: Praeger Security International, 2009), pp. 38–59.

[16] One of the reasons for Malaysia and Indonesia to turn towards Russia and other former Warsaw Pact states is the fact that these states have been willing to conduct such trade on part cash, part barter terms that suit the tighter financial conditions that both Southeast Asian states face.

[17] For a recent study of emerging technologies from the United States, see: Michael O'Hanlon, "Forecasting Change in Military Technology, 2020–2040," Brookings Institution, September 2018 (https://www.brookings.edu/research/forecasting-change-in-military-technology-2020–2040/).

[18] Stephen Biddle, *Military Power: Explaining Victory and Defeat in Modern Battle* (Princeton and Oxford: Princeton University Press, 2004). Biddle describes the modern system as "a tightly interrelated complex of cover, concealment, dispersion, suppression, small-unit independent manoeuvre, and combined arms at the tactical level, and depth, reserves, and differential concentration at the operational level."

[19] Ron Matthews, *European Arms Collaboration*, Harwood Academic Press (1992), ch. 3; Ron Matthews, "International Arms Collaboration: The Case of Eurofighter," *International Journal of Aerospace Management*, vol. 1, no. 1 February 2001, pp. 73–9.

[20] The Voyager spacecraft had about 70 kilobytes of memory, in comparison to the iPhone 5, which has 16 gigabytes of memory; iPhone 5 therefore as 240,000 times more memory than Voyager. See: https://www.nasa.gov/mission_pages/voyager/multimedia/vgrmemory.html#.X9q28C0RphA.

[21] See, for instance: Bernard Fook Weng Loo (ed.), *Transformation and Military Operations* (London and New York: Routledge, 2009).

[22] *Military Transformation: A Strategic Approach* (Washington, DC: Department of Defense, United States of America, 2003), pp. 20–27.

[23] Kapil Kak, "Revolution in Military Affairs-An Appraisal," *Strategic Analysis*, April 2000.

[24] *Joint Vision* 2020 (accessed online: http://www.dtic.mil/jointvision/vpub2.htm., 1 May 2008), p. 6. Emphasis own.

[25] *Ibid.*

[26] *Ibid.*, p. 10.

[27] Robert W. Chandler, *The New Face of War: Weapons of Mass Destruction and the Revitalisation of America's Transoceanic Military Strategy* (McLean, VA: AMCODA Press, 1998).

[28] Antulio J. Echevarria II, *Toward an American Way of War* (Carlisle PA: Strategic Studies Institute, Army War College, 2004), p. vi.

[29] Arthur K. Cebrowski and Thomas P.M. Barnett, "The American Way of War," in *Transformation Trends*, 13 January 2003, p. 1.

[30] For instance, see Victor Davis Hanson, *Culture and Carnage* (New York: Doubleday, 2001), p. 97.

[31] Bernard Fook Weng Loo, "Decisive Battle, Victory, and the Revolution in Military Affairs," *Journal of Strategic Studies*, vol. 32, no. 2, April 2009,

pp. 189–211; Bernard Fook Weng Loo, "The Challenges Facing 21ˢᵗ Century Military Modernization," *PRISM The Journal of Complex Operations*, vol. 8, no. 3, January 2020, pp. 146–157.

[32] Thomas G. Manhken and James R. FitzSimonds, "Tread-Heads or Technophiles? Army Officer Attitudes towards Transformation" in *Parameters*, vol. XXXIV, no. 2, Summer 2004, pp. 52–72. In this study, "military transformation" is understood as the RMA operationalised as official policy.

[33] Mark Beeson and Alex J. Bellamy, *Securing Southeast Asia: The Politics of Security Sector Reform* (London & New York: Routledge, 2008), p. 59.

[34] Tim Huxley, "The ASEAN States' Defence Policies 1975–81: Military Responses to IndoChina," *Strategic and Defence Studies Centre Working Paper* No. 88 (Canberra: Australia National University, 1984), p. 11.

[35] Cited in Robin Laird and Holger Mey, "The Revolution in Military Affairs: Allied Perspectives," *McNair Paper* 60 (Washington, DC: National Defense University, 1999).

[36] For those who insist that a new security agenda has emerged, that "traditional" wars have become less relevant, see, among others: Mary Kaldor, *New and Old Wars: Organized Violence in a Global Era* (Cambridge: Polity Press, 1999); Thomas X. Hammes, *The Sling and The Stone: On War in the 21ˢᵗ Century* (St. Paul, MN: Zenith Press, 2004); and Bruce Berkowitz, *The New Face of War: How War Will Be Fought in the 21ˢᵗ Century* (New York and London: The Free Press, 2003). A counter-argument to these scholars is Colin S. Gray, *Another Bloody Century: Future Warfare* (London: Weidenfeld & Nicolson, 2005).

[37] Moises Naim, "Five Wars of Globalisation," *Foreign Policy*, January/February, 2003, pp. 29–36.

[38] See, for instance: Kalevi J. Holsti, *The Decline of Interstate War, or The Waning of Major War* (London: Routledge, 2006); and Thomas Szayna *et al.*, *What are the Trends in Armed Conflict, and What Do They Mean for U.S. Defense Policy?* (Santa Monica: RAND, 2017); Kendra Dupuy and Siri Aas Ruustad, *Trends in Armed Conflict, 1946–2017* (Oslo: Peace Research Institute, 2018). Rupert Smith has gone so far as to argue, "war as cognitively known to most non-combatants, war as battle in a field between men and machinery, war as a massive deciding event in a dispute in international affairs: such war no longer exists." See: Rupert Smith, *The Utility of Force: The Art of War in the Modern World* (New York: Alfred A. Knopf, 2007), p. 3.

[39] For the former, see: https://ourworldindata.org/terrorism; for the latter, see: https://world101.cfr.org/global-era-issues/terrorism/terrorism-numbers.

[40] Hammes, *The Sling and the Stone*, p. 260.

[41] Michael W. Doyle, "Discovering the Limits and Potential of Peacekeeping," in Olara A. Otunnu and Michael W. Doyle (eds.), *Peacemaking and Peacekeeping for the New Century* (Oxford: Rowman and Littlefield, 1998), pp. 8–12.

[42] Charles Moskos *et al.* (eds.), *The Postmodern Military: Armed Forces after the Cold War* (Oxford: Oxford University Press, 2000).

[43] Bruce Berkowitz, *The New Face of War: How War will be fought in the 21st Century* (New York: The Free Press, 2003).

[44] See Kumar Ramakrishna and Bernard Loo, "The US Military and Non-Conventional Warfare: Is Firepower Cheaper than Manpower?," *IDSS Commentary* No. 34, 21 July 2005; and Bernard Loo, "The Military and Counter-Terrorism," *IDSS Commentary* No. 89, 8 December 2005.

[45] Jon Grevatt, "Indonesian defence minister reaffirms procurement priorities," *Jane's Defence Industry*, 16 May 2008.

[46] Adrian David, "Need for more air power to protect assets, says forces chief," *The New Straits Times*, 16 September 2009.

[47] "High-tech gadgets for straits security," *The Star*, 21 March 2009.

[48] Bernard F.W. Loo, "From Poisoned Shrimp to Porcupine to Dolphin: Cultural and Geographic Perspectives of the Evolution of Singapore's Strategic Posture," in Amitav Acharya and Lee Lai To (eds.), *Asia in the New Millennium: APISA First Congress Proceedings, 27–30 November 2003* (Singapore: Marshall Cavendish Academic, 2004), pp. 352–375; Norman Vasu and Bernard F.W. Loo, "National Security and Singapore: An Assessment," in Terence Chong (ed.), *Management of Success* (Singapore: Institute of Southeast Asian Studies, 2010). Updated and republished in Barry Desker and Ang Cheng Guan (eds.), *Perspectives on the Security of Singapore: The First 50 Years* (Singapore: World Scientific and Imperial College Press, 2015), pp. 21–43.

[49] Bernard F.W. Loo, "The Management of Military Change: The Case of the Singapore Armed Forces," in Jo Inge Bekkevold, Ian Bowers and Michael Raska (eds.), *Security, Strategy and Military Change in the 21st Century* (London and New York: Routledge, 2015), pp. 70–88.

https://doi.org/10.1142/9789811244292_0006

Chapter 5

The Changing Character of War in the 21st Century Challenges for Strategic Planning

Bernard F.W. Loo

*Senior Fellow, Institute of Defence and Strategic Studies,
S Rajaratnam School of Internal Studies,
Nanyang Technological University, Singapore*

isfwloo@ntu.edu.sg

Between 1979 and 1989, there were an estimated 5,000–20,000 non-Afghan nationals—from the Middle East and North Africa, Southeast Asia, North America and Europe—who fought alongside the Afghan Mujahideen against the former Soviet Union. There were an estimated 2,000 foreign fighters in Bosnia between 1992–1995. In the war in Iraq (2003–2011), there were an estimated 5000 foreign fighters that came principally from other Arab countries, but also included European and American nationals as well.[1] A study, published in December 2015, estimates that there are between 27,000–31,000 foreign fighters, coming from at least 86 countries, that have moved to Syria and Iraq to join the Islamic State.[2] Indeed, as Steven Metz wrote,

> the Islamic State has woven together a dangerous network, this one composed of fat-cat Gulf funders, angry young Western Muslims

struggling with inner demons, local Sunni Arabs angered by repression from the governments in Damascus and Baghdad, violence-obsessed jihadists from across the Islamic world and former Baathists still bitter over losing power.[3]

On 22 March 2017, Khalid Masood drove a rented vehicle onto the sidewalk along Westminster Bridge in London, near the Houses of Parliament. He killed two individuals instantly, and a third subsequently died from injuries sustained by the attack. Khalid Masood subsequently escaped on foot and fatally stabbed a police officer who had attempted to apprehend him, before he was shot and killed by another police officer. On 12 August 2017, in Charlottesville, North Carolina, James Field drove a car into a crowd of people who were part of an anti-right wing demonstration; one woman was killed.

The focus of this chapter is on how the technological zeitgeist of the early 21st Century—in particular, the ubiquity of information and communications technologies—makes both foreign fighters and so-called lone-wolf actors such as Khalid Masood and James Field possible, thereby changing the character of war. The processes through which these foreign fighters and lone wolves are mobilized potentially constitute a defining characteristic of wars in the 21st Century. Following from this, if the character of war in the 21st Century is different from its predecessors, correspondingly, the metaphors of war that strategists and scholars use to understand the phenomenon of war need to change.

Why does the issue of metaphors of war matter? Carl von Clausewitz, probably the most important lodestone of strategic thinking, used the metaphors of war as a duel or as a wrestling match: "War is nothing but a **duel** on a larger scale. Countless duels go to make up the war, but **a picture of it as a whole can be formed by imagining a pair of wrestlers** (emphasis mine)."[4] How war is therefore understood subsequently matters because it then begins to condition the kinds of strategies that ought or ought not to be adopted in waging these wars. Correspondingly, the metaphors of war—the images that scholars and practitioners of war use to understand otherwise complex phenomena such as war—ought to accurately reflect the character of war. But can this Clausewitzian metaphor be used to accurately understand the wars of the 21st Century? This chapter will propose that a more accurate metaphor of war today might be a street brawl, the combatants more accurately framed as a kind of "flash mob."

This chapter begins with a brief discussion of what is arguably the most important defining characteristic of war in the 21ˢᵗ Century, namely the ubiquity and pervasiveness of information and our ability to access this information, and how this leads to the subsequent transformation of the politics of individual and national identities. It is in this context that we can study the issue of foreign fighters, and the processes through which these foreign fighters and lone wolves are mobilized. Finally, the attention shifts to the implications of these changes for the character of war in the 21ˢᵗ Century, through the concept of metaphors; how the Clausewitzian metaphors of war, its features and its implications thereafter for waging war may no longer apply; and its implications for the formulation of strategy.

Information Proliferation and the Politics of Identity

Two main features of this technological character of the 21ˢᵗ Century are germane: the generation of data, and the capacity to communicate and transmit this data almost instantaneously around the world. In the first instance, data has grown exponentially in the 21ˢᵗ Century. *ScienceDaily* estimated in 2013 that 90% of the world's data was generated between 2011 and 2013; much of this "data" was highly unstructured, comprising of photo imagery and videos, sound files and text.[5] An International Data Corporation study published in 2011 predicted that the amount of information that would be generated in that year alone amounted to 1.8 zettabytes—that is 1,800,000,000,000,000,000,000 bytes—and that between 2011 and 2020, the amount of global data will increase by 50 times.[6] One study reports that the data that was generated in 2011 actually amounted to 5 zettabytes, and that 41 zettabytes of data were generated in 2019.[7] The World Economic Forum predicted that 44 zettabytes of information will be created in 2020.[8]

Furthermore, this information is very easily accessed. Before the advent of broadband and optic fiber connections, the information that was available on the Internet could be accessed, but it was difficult and slow; that has all changed now. Between 2000 and 2014, the number of Internet users increased by 741%,[9] amounting to over 3 billion Internet users.[10] Over the last decade, wireless broadband is now widely available. This has had a further impact on our ability to access the information that is being generated. Finally, the introduction of social media—the most famous being Facebook—not only allows us to access information, it allows us to disseminate information that we create to potentially a world-wide audience. As of July 2020, there are 4.66 billion internet users, 4.14 billion active social media users, of which 4.08 billion are active on mobile devices.[11] Facebook has over 2.7 billion users, Whatsapp and YouTube have 2 billion users, Instagram has 1.16.[12]

As noted earlier, this chapter focuses on how this easy access to information changes the character of war in the 21st Century. To understand this, we need to examine how our ability to gain access to ever-increasing amounts of information has an impact on our ability to understand the information we are gaining access to.

To cognitive science, there are potentially negative consequences to this kind of communicative immediacy. As Daniel Levitin noted, "Because it is limited in characters, it discourages thoughtful discussion or any level of detail."[13] Maggie Jackson further warned in 2009 that the capacity to access so much information so quickly has a potentially adverse consequence on the capacity to think critically: "Half of college students can't judge the objectivity of a website."[14] Merlin Donald has argued that the digital age in which we live represents a

> **cognitive** (emphasis original) revolution. The new media are aimed at the mind. They are interconnected with the sense organs. They aim their sophisticated, carefully engineered messages directly at the memory systems of the brain. They actually restructure memory, changing both the storage and retrieval systems we depend upon, and they are addressed directly to the source of our experience, and aimed at consciousness itself.[15]

The digital information revolution has resulted in a situation that encourages a "shallow mode of learning characterized by quick scanning, reduced contemplation and memory consolidation."[16] The sheer volume of information that is generated, and the immediacy with which consumers can access this information, inhibits deep comprehension of any issue; rather, any given issue tends to be understood in simplistic, binary terms. Carefully nuanced comprehension of complex issues, under such conditions of information overload, becomes difficult, if not impossible. Advertisers understand this cognitive condition, and seek to actively exploit it, to persuade consumers to purchase goods that the latter does not necessarily require.

To a number of strategic actors, this condition of "shallow learning" creates the condition upon which the next issue, the transformation of politics at the levels of the state, the nation, and individual human identity, can occur. Furthermore, this transformation forms the context, the "necessary condition for the rise of foreign fighters."[17] As Jack David Eller has argued, the apparently ethnic nature of conflicts that have arguably characterized the post-Cold War world order are themselves inventions of the time, based on identities that are the result of processes that may be understood to be wholly artificial in nature: "States, even when they are real, are not 'natural' or 'given' but are generally products of a historical and political (often a military or colonial) process, made and broken and remade as groups and/or individuals vie for control over territory."[18]

The most famous proponent of the idea that nation is an "imagined community" is, of course, Benedict Anderson; as he argued, the "nation"

is **imagined** because the members of even the smallest nation will never know most of their fellow-members, meet them, or even hear of them, yet in the minds of each lives the image of their communion ... it is imagined as a **community**, because, regardless of the actual inequality and exploitation that may prevail in each, the nation is always conceived as a deep, horizontal comradeship (emphases original).[19]

Anderson later argued that it is precisely "advances in communications technology, especially radio and television [that] can conjure up the imagined community to illiterates and populations with different mother-tongues."[20] Political elites understand this, and therefore deliberately leverage on these technologies to create the idea of the nation—the imagined community of otherwise disparate actors and individuals.

Identities are, in other words, malleable. For the individuals who make up this "imagined community," however, the "nation" ought to be understood as merely one possible identity, certainly not the **only** identity. This notion that an individual can have multiple identities is an idea found in the three major strands of identity theory: internalized role-identity theories; cultural and situational context theories; and collective identity theories.[21] In the first two strands, identities "serve as an organizational force, binding us to those with commonalities of interest and providing a social glue that can serve as a foundation for mobilizing joint action."[22] As Margaret Somers argues, individual identities are socially constituted.[23] Identities are therefore not fixed or essentialist in nature, but rather "are constituted and reconstituted over time."[24] Somers further argues "that social life is itself *storied* and that narrative is an *ontological condition of social life* (emphasis original) [and] that people construct identities (however multiple and changing)."[25] The third strand of identity theory, collective identity theory, does not preclude the two preceding strands; rather, they "should not be seen as a third type of identity as much as an attempt by this literature to highlight another set of identity-related dynamics at the group level."[26] This tendency towards multiple collective identities is accentuated in the technological conditions of the 21st Century discussed earlier.[27]

It is the perception of fraternity with the travails and suffering of far-flung communities that enables people from elsewhere to be prepared to make the ultimate sacrifice for those communities.[28] The perceived fraternity is precisely that—perceived, rather than real—and it is driven by the shallow learning that arises out of the technological zeitgeist of the 21st Century; it is precisely "global communications [that] has greatly increased the visibility of war as well as the sense of solidarity with strangers."[29] The malleability of identity means that, under the technological conditions identified earlier, otherwise disconnected people can come to identify with the political cause of other groups found elsewhere in the world. It is that collective identity that acts as an "intervening causal mechanism in situations of 'objective' social change."[30] Precisely because individual identity can be understood as a "context-dependent cognitive representation or process,"[31] it is possible to leverage on conditions of shallow learning to create "an intrinsic 'tribal' sense of a unique historical and ethnic identity."[32] This is a variation of the triangulated or tripolar relationship that Simon and Klandermans posit: two main antagonists, and a general public that at least one of the antagonists may actively seek to enlist for its own interests.[33] Alternatively, this may point to self-categorization theory—where people perceive themselves as both individuals and as members of a social group.[34]

The Islamic State, for instance, understood this cognitive condition, and sought to exploit this condition through a well-organized media bureaucracy that generated an output—whether in the form of tweets, web pages, or on-line videos—that was characteristically slick, professional, and had good production values.[35] It employed the same psychological tricks as advertisers and marketing agencies, to establish connections with peoples from around the world to its agenda. The success of the Islamic State's media campaigns can be seen in the number of foreign fighters who went to Syria and Iraq. However, the Islamic State was also able to craft success at a more insidious level: Thomas Rid and Marc Hecker provide the example of Younis Tsouli, a cyber-terrorist of Moroccan origins but residing in London, who "complained to another online-jihadist [that] 'my heart is in Iraq.'"[36] A politicized collective identity begins to emerge amongst otherwise disparate and widely distributed peoples of different historical, socio-cultural and political backgrounds. This emergence may be due to an awareness that existing grievances in one part of the world are shared by peoples in other regions of the world, but nevertheless connected emotionally for whatever reason real or imagined.[37] It is this process of politicized collective identity that lays the foundations for the increasing strategic salience of so-called foreign fighters in the wars of the 21ˢᵗ Century. This issue will be examined next.

Foreign Fighters and Lone Wolves as New Strategic Actors

Thomas Heghammer defines foreign fighters as actors who: join an existing armed conflict; are not citizens of the state or states involved in this armed conflict; lack kinship links with the various warring factions; and are unpaid by the state or warring faction that they have joined.[38] Finally, these foreign fighters may enjoy "tacit, but not active, state support."[39] A key issue is the extent to which these foreign fighters are connected in some form or another to any of the warring factions involved in the given armed conflict. David Malet argues that ethnic or other connections may exist between the domestic and foreign fighters, but suggests also that such connections are not absolutely necessary.[40]

During the Spanish Civil War (1936–1939), there were a reported 40,000 international volunteers.[41] Amongst these international volunteers was the so-called Abraham Lincoln Brigade, comprising 2,800 American volunteers.[42] What united these American volunteers was a shared antipathy towards Fascism as an ideology, but what also mattered was how that Fascism related to their individual backgrounds. For some Americans, it was the threat that Fascism posed to democracies worldwide. For African Americans, it was the inherent racist character of Fascism to which they objected. For Jewish Americans, it was the specific element of Anti-Semitism that drove their decisions to volunteer.[43] A significant issue is the commitment of these American volunteers to their cause. After the Spanish Civil War ended, these American volunteers returned home, only to continue their struggle against the Fascist Franco regime, despite the legal and political challenges they faced back home.[44] At the same time, however, the presence of the international brigades was not merely "a simple and straightforward contest between democracy and fascism."[45] The Comintern's involvement, for instance, reflected in part the Kremlin's Popular Front strategy, which was to "secure as wide a spectrum of support as possible within the bourgeois democracies for foreign policies that would, in effect, support those of the USSR."[46]

The Soviet Union invaded Afghanistan in December 1979, and very quickly had to face an opposition that comprised a fairly large number of foreign fighters drawn from the Middle East, South Asia and Southeast Asia. These foreign fighters were, according to Heghammer, part of an emerging pan-Islamic identity discourse mobilized through a "massive propaganda effort" based on print media such as globally distributed magazines published in numerous languages as well as an extensive network of Islamic charity organizations.[47] There is also evidence to suggest that Arab involvement on the side of the mujahidin was due in part to the fact that a number of Afghan mujahidin leaders had studied in Islamic universities in Egypt in the 1960s.[48] Certainly there were Afghan warlords who had welcomed fighters from the Middle East.[49] Who were these people who traveled to Afghanistan to fight the Soviet invaders? One report indicates that they were "highly motivated and prepared to die for the jihad," that they had "spent their own money to volunteer," and about 30% of them were "criminals or outlaws in their own countries."[50]

More recently, in the current conflict in Iraq and Syria, the Islamic State has been able to call upon a fairly significant number of foreign fighters drawn from around the world. One study found that many foreign fighters were mobilized not through official channels, but by information from so-called disseminators—"unaffiliated but broadly sympathetic individuals who can sometimes appear to offer moral and intellectual support to jihadist opposition groups."[51] Foreign fighters have also joined various groups fighting against the Islamic State.[52]

The presence of foreign fighters in any armed conflict points to at least two processes at work, namely active recruitment and self-mobilization. In the first instance, by utilizing modern communications technologies—an issue that I will discuss later in this study—one side involved in an on-going armed conflict can actively recruit foreign fighters. Indeed, as Rid and Hecker argue, the ability to generate global public support by utilizing "commoditized telecommunications to build up and protect their own will and their own capabilities"[53] is key to the strategy of insurgent groups such as Al Qaeda, Hezbollah and the Taliban.[54] Such insurgent groups will use communications media to build their political bases and constituencies, train and recruit operatives from sympathetic individuals.[55] As Malet argues, recruitment of foreign fighters tends to occur through "social networks of the transnational communities, and potential recruits are generally closely connected to these identity sub-groups rather than to their wider national society."[56]

In the second instance, foreign fighters may emerge as a result of a process of self-mobilization. Malet argues that the presence of foreign fighters in at least one side of an existing armed conflict therefore suggests the existence of a shared identity—"whether religious, ideological, or nationalist"—between the warring side and the foreign fighters.[57] It is therefore possible that foreign fighters exist not because of active recruitment by a combatant side, but rather because these foreign fighters individually believe themselves to be connected to that side. This sense of connection—even in the absence of kinship or ethnic ties—exists because identity is a potentially malleable concept, an issue I will explore in greater depth later in this study. Furthermore, these reasons are not mutually exclusive; foreign fighters may have more than one reason for joining an existing armed conflict.[58]

The preceding brief survey of foreign fighters in earlier conflicts points to two issues that will be examined in greater detail later. One, the willingness of American volunteers in the Spanish Civil War to continue the struggle, to persist in the face of military defeat, points to the importance of the will to persist in strategy and war.[59] Two, the cases mentioned above also highlight how foreign fighters can present themselves in an on-going armed conflict for multiple reasons. If foreign fighters are not a 21ˢᵗ Century phenomenon, it then begs the question of why the presence of foreign fighters in 21ˢᵗ Century conflicts might signify a change in the character of wars, thus requiring new metaphors of war.

To reiterate, it is because of the perception of community with people directly involved in existing conflicts that makes otherwise unconnected people from disparate regions of the world decide that they need to "do something"—in this case, travelling to the conflict zone to fight alongside their perceived community. This phenomenon of disparate peoples travelling great distances to join an on-going armed conflict resembles the early 21st Century phenomenon of the flash mob: whereby groups of otherwise unconnected people are mobilized, usually through some form of mobile mass communication, to turn up at a specified location and time to perform a particular action.[60] The originator of the flash mob, Bill Wasik, claims that he had sent an email to over sixty friends and acquaintances on 27 May 2003 inviting them to participate in a "project that creates an inexplicable mob of people" and to "forward this [message] to other people you know who might like to join."[61]

Similarly, the presence of foreign fighters in the conflicts mentioned earlier points to how people from elsewhere can become mobilized through different communications media to participate in a conflict far away from their homes. In the case of Americans participating as private citizens in the two World Wars or the Spanish Civil War, these were the result individuals being self-mobilized by the information they were receiving. The Islamic State, however, has consciously exploited communications media to broadcast its message to a potentially global audience, and actively seeks to recruit other nationals—whether as foreign fighters or as non-combatants performing essential services. Indeed, the list of non-combatant roles that ISIS has sought to attract includes, amongst others, medical professionals, legal professionals, women who are inspired to become wives of ISIS's soldiers, as ISIS seeks to create a new society.[62] People subsequently respond positively to these calls, and journey to these conflict zones as a result. This "flash mob" is not necessarily a unitary actor, let alone a structurally coherent organizational actor. If the above analysis is true, as need to rethink our analogies and metaphors with which we understand war, warfare and combatants.

Towards a New Metaphor of War in the 21st Century

Metaphors are, as cognitive psychologist George Lakoff has argued, "the way we conceptualize one mental domain in terms of another."[63] In other words, metaphors use images and analogies to provide us with a means of perceiving and understanding otherwise complex phenomena. Cognitive science shows that human reasoning occurs "in terms of frames and metaphors" and that these metaphors provide us with frames around which we organize and comprehend new, even potentially discordant, information.[64] Metaphors provide us with frameworks for "understanding a new domain or for restructuring the understanding of a familiar domain."[65]

It is, of course, possible to argue that war as a phenomenon can be understood even by non-professionals, without having to resort to metaphors; Rupert Smith, for instance, argued that lay persons intuitively understand war as a "battle in a field between men and machinery, war as a massive deciding event in a dispute in international affairs ..."[66] Nevertheless, war is a highly complex phenomenon. In its totality, it comprises many components: pre-war perceptions of putative adversaries and the construction of a military organization that is capable of maintaining a more or less peaceful security environment through the exercise of deterrence postures; a regular modernization process that ensures that the national military instrument is kept as up-to-date and capable as is economically supportable; the mobilization of military resources in times of inter-state crises; the political decision to resort to the use of military force to achieve political end-states that policy-maker have deemed to be absolutely essential to their national interest; the actual conduct of military operations in an action-reaction dynamic between the combatant parties. All of these, and more, come together to form the phenomenon of war. Metaphors are therefore, as noted earlier, absolutely essential for policy makers and strategic planners to understand what exactly it is that they are doing, or about to do.

As an illustration, the United States' debate over its policy options regarding the Kosovo conflict was conducted as a "metaphor war," a contest for "interpretive dominance ... that is, the widespread acceptance of one's own characterization of a particular issue."[67] If Kosovo was understood as another holocaust, this would demand some form of military intervention; if understood as a Vietnam, this would generate "countervailing pressures against intervention."[68]

In an earlier era, United States policy-makers and strategic planners employed three historical episodes—the Munich Conference of 30 September 1938, the battle of Dien Bien Phu and the collapse of France's Indochinese empire, and the Korean War—to argue for or against the proposition that the United States needed to escalate its commitment to the defense of the former South Vietnam.[69] If policy debates over decisions to go to war employ the use of metaphors, metaphors are also essential for how policy-makers and strategic planners understand the war they are then engaged in.

As noted earlier, Clausewitz posited war as an essentially two-sided contest, "an act of force to compel our enemy to do our will" and a "collision of two living forces."[70] The Clausewitzian metaphor therefore posits war as a phenomenon that involves two essentially monolithic, unitary actors. If war is framed as a "duel" and its combatants as "a pair of wrestlers," war becomes associated with the cultural norms and formal rules that are associated with dueling and wrestling respectively.[71] Clausewitz's metaphor of war, then, exists within a particular context, shaped by international norms and laws concerning the use of force, which is an issue that will be examined in greater detail later.

Furthermore, if war is a clash of wills, as I have argued elsewhere, it is possible to deduce that the real "battlefield" is always the opponent's will to resist.[72] At the operational and tactical levels, it is a matter of breaking the morale of the opposing troops. At the strategic level, will is manifested typically in the political leadership and/or the hearts and minds of the population of the sides involved, and their respective willingness to endure the otherwise unendurable. If war is a duel between two unitary actors, the appropriate strategy that brings about strategic success is one that compels one unitary actor to accede to its opponent's political will. However, even in a war between two unitary actors, collapsing the will of the opponent is already a potentially problematic task.[73] If war is conceptualized as a street brawl involving many actors coming together as a flash mob, then strategic success can only come about once the will of the individual actors that comprises this flash mob have been defeated. This is an issue that will be examined later.

What evidence is there that these Clausewitzian metaphors continue to shape the manner in which strategic planners and policy makers think about war? The evidence can be found in a number of sources.

One key source is the literature on strategy and war, in particular literature on recent wars or armed conflicts and the challenges that these recent wars pose to strategy and strategic planners. A number of important texts on wars in the 21st Century—for instance, David Kilcullen's *The Accidental Guerrilla*, and Thomas Rid and Marc Hecker's *War 2.0*—arguably continue to posit wars in the 21st Century as essentially two-sided affairs. Kilcullen's "accidental guerrillas" are "widely spaced and disparate microactors [who] can aggregate their effects, to achieve outcomes disproportionate to the size and sophistication of their networks."[74] This posits a more or less monolithic actor (in his case, Al Qaeda) that is able to exploit the localized backlash against Western intervention "in order to generate support for its *takfiri* agenda."[75] The "accidental guerrilla" is a strategic actor who fights "us not because they seek our destruction but because they believe we seek theirs, a belief in which they are encouraged by a cynical, manipulative clique of *takfiri* terrorists."[76] Rid and Hecker focus their analysis on Hezbollah, the Taliban and Al Qaeda, and how each group has a more or less coherent strategy that targets public opinion, to deny political legitimacy to their opponents while enhancing their bases of financial, moral and material support.[77]

Even the concept of hybrid warfare tends to posit war between two more or less monolithic actors, inasmuch as at least one side deploys a number of non-traditional elements—such as cyber attacks perpetrated by so-called civilians—as part of its overall strategy. Frank Hoffman describes hybrid warfare as a situation where "the **adversary** (emphasis mine) will most likely present unique combinational or **hybrid** (emphasis original) threats."[78] This adversary, whether a state or non-state actor, will "exploit access to modern military capabilities … as well as promote protracted insurgencies that employ ambushes, improvised explosive devises and coercive assassinations."[79] Irregular tactics are therefore not merely the instrument of the weak non-state actors, but rather are the "tactics of the **smart and nimble** (emphasis original)."[80]

A second source can be found in the force structures, operating concepts and doctrines, and training systems that armed forces use to prepare themselves for their respective missions. For instance, the United States' *2014 Quadrennial Defense Review* (QDR) portrayed a future security environment that includes "Global connections … resulting in greater interaction between states, non-state entities, and private citizens,"[81] but nevertheless seeks to "project power and win decisively"[82] by developing new assets such as F-35 fighter aircraft, stealthy long-range strike aircraft, a new class of ballistic missile submarines. Such capabilities are intuitively strategically necessary in a war that reflects the Clausewitzian metaphor of an essentially two-sided war.[83] Such capabilities are arguably less strategically viable in a war conceptualized as a street brawl involving individual actors who come together as a flash mob. Similarly, a 1994 British Ministry of Defence publication, *Wider Peacekeeping*, foresaw armed conflict involving "'numerous parties' who thrived in a general 'absence' of law and order," although military scholars subsequently argued for the need to separate "peacekeeping from other military operations."[84] As the military deployments to Afghanistan and Iraq started to wind down, the military organizations of Britain and the United States started to move their respective training regimes back to conventional military operations, predicated on a single, organizationally coherent, putative adversary force.[85] While such military capabilities might be potentially applicable in a war-as-duel, its applicability, as I will suggest later, to war-as-street brawl is at best questionable.

But what if this Clausewitzian metaphor of war no longer applies to the wars that states are likely to face in the 21st Century? Clausewitz himself argued that "every age has its own kind of war, its own limiting conditions, and its own peculiar preconceptions. Each period would have held to its own theory of war..."[86] If this is true, then it is not implausible that the 21st Century has its own theory of war, and correspondingly its own metaphor of war. The "opponent" a state is likely to face in a 21st Century war is unlikely to be a unitary actor, but rather a loose collection of strategic actors, each motivated by objectives peculiar to that strategic actor, and the only thing that may unite these otherwise disparate strategic actors is their shared antipathy to that state.

The question is how and why disparate groups of people widely distributed across the globe might come to believe that an existing conflict elsewhere might pose a threat to their respective values such that these groups of people are willing to travel, at least in some cases, great distances to participate in these wars. The answer to this question can be found in the changing character of war, which will be examined next.

The Implications for Strategy

This means that an inter-state war can quickly spiral out of control as foreign fighters, whether actively mobilized by either combatant state or self-mobilized, begin to flood into the conflict. What may have started out as a Clausewitzian duel eventually transformed into something more akin to a street brawl. A street brawl might begin as a straightforward fight between two actors, but as the dynamic of violence begins to spiral upwards and outwards, otherwise uninvolved people begin to join in the fight for their own particular reasons. For instance, the 1992 Los Angeles riots started as demonstrations against the acquittal of police officers on trial over the brutal arrest of Rodney King, but soon spiraled out of control into acts of looting and arson as well. Similarly, the "intrusion" of foreign fighters from disparate origins and possibly different motivations creates the possibility of the objectives and the nature and character of the original fight changing. In both instances, the psychology of the mob may take over the strategy of the combatant party.[87] A brief examination of mob sociology and psychology provides some insights into the strategic behavior of these strategic "flash mobs."

The crowd or the mob is intimately linked to the emergence of industrialization and the growth of major urban centers in Europe and North America, for two reasons.[88] One, it led to a breakdown in traditional religious and familial hierarchies, which itself "led to a level of rootlessness and mindlessness which made the mass prey to anarchic impulses, to passing fads and to unscrupulous agitators."[89] Two, urbanization facilitated social networking and mass mobilization.

Crowd or mob behavior can be portrayed in both negative and positive terms. In the first instance, some crowd theorists portray crowd behavior as essentially negative, as a kind of contagion: "they are unable to resist any passing idea or, more particularly and because the intellect is all but obliterated, any passing emotion."[90] As Schweingruber argued,

> The individuals in the crowd, who are already "preconditioned" and have "built-up" frustrations, "mutter" and "mill about like a herd of cattle," "jostling" and "name-calling." During the milling, the crowd members spread rumours, which are crucial in the development of a mob. There is a "spiral of stimulation" ... The individual is absorbed into the crowd and is controlled by the "crowd mind." This leads to violent and destructive behaviour. The target and details of this behaviour can be influenced by leaders or agitators. These agitators can cause the precipitating incident that creates the mob, or the incident can be triggered by unjustified actions of the police.[91]

Furthermore, as Gustave Le Bon, arguably the dominant scholar of crowd theory, suggested, crowds allow individuals to "lose all sense of self and all sense of responsibility [while gaining] a sentiment of invincible power due to their numbers."[92] Crowd behavior borders on barbarism. Even if contemporary theories of crowd behavior have attempted to move away from Le Bon, they nevertheless retain notions of deindividuation, impulse and chaos as at least potential if not necessary elements of the crowd.

It is precisely the transformations of identity and nationhood, facilitated by the proliferation of information and the consequent shallow cognition, that allow different ethnicities around the world to be mobilized by something happening somewhere to want to do something about it. In the case of Afghanistan, Iraq and Syria, it is some Muslims in different countries who have been mobilized to want to go there and do something about it; the fact that not all Muslims around the world are following this example clearly demonstrates it is not the belief in the faith that explains this phenomenon. If nations are "imagined communities," then, as Homi Bhabha has argued, "Nations, like narratives, lose their origins in the myths of time and only fully realize their horizons in the mind's eye."[93] In other words, it is the concept of the nation that allows for otherwise disparate people located throughout the world to feel connected to a particular geopolitical situation. As Michael Ignatieff has argued,

> As a cultural ideal, nationalism is the claim that while men and women have many identities, it is the nation that provides them with their primary form of belonging. As a moral ideal, nationalism is an ethic of heroic sacrifice, justifying the use of violence in the defense of one's nation against enemies, internal or external.[94]

What does this argument therefore suggest about the issue of victory, or strategic success, in the wars of the 21st Century? As suggested earlier, securing strategic success in a war framed as a Clausewitzian duel or wrestling match is already difficult enough. Given that war is ultimately a clash of wills, strategic success is determined by the will of the opponent to persist. Indeed, the essential condition for strategic success is not strategic coherence on one's side, but rather the lack of will on the other. Regardless of the metaphor of war, defeat goes to the side that decides its desired political end-states are not worth the costs incurred. Conversely, strategic success is conferred upon the side that refuses to give up, the side that decides that its desired political end-states are worth any price.

But if war in the 21st Century is framed as a street brawl, then does this mean that everyone involved in the street brawl has to concede defeat before strategic success can be claimed? While every foreign fighter who has gone to Syria and Iraq to join the Islamic State may share in the latter's geopolitical vision, it is very likely that these foreign fighters also have personal motivations separate from that of the geopolitical vision of the Islamic State. Precisely because this "flash mob" is neither a unitary actor nor organizationally coherent, defeating the 'flash mob' will require persuading every individual to accept defeat.

Of course, it is important, even necessary, that one's strategy is coherent in the first place. Policy-makers need to develop a coherent vision of the desired political outcomes, and then give sufficient guidance to military planners to derive the strategic outcomes that military force can then aim towards. In other words, the military objectives must be consonant with the political objectives. The ability to translate the political objectives into plans that provide guidelines for the conduct of military operations against clearly identified targets and objectives is therefore essential to success in war. Strategy, after all, is essentially a discourse between military planners and policy makers. Fighting any battle is simply about creating effects that facilitate the attainment of broader military and ultimately political objectives. However, military organizations that face a "flash mob" are operating at a potential disadvantage. The top echelons of military organizations are almost certainly less familiar with, and consequently lack an intuitive grasp of, contemporary communications media. They are not so-called digital natives, unlike the younger generation of military officers, who are most likely to intuitively understand the technological conditions of war in the 21ˢᵗ Century.[95] Of course, military planners and policy makers also ought to ask questions about the probabilities of unintended effects, especially counter-productive effects, even if this is a difficult task at best.[96] As the wars in Korea, Vietnam, Afghanistan and Iraq seem to demonstrate, the eventual outcomes are not necessarily driven by battle outcomes. If, as earlier suggested, it is difficult to persuade the opponent—assuming the opponent to be a coherent unitary actor—in a *zweikampf* to accept defeat, then surely it becomes even harder in a street brawl to persuade all the various actors to accept defeat.

A further complication for strategy lies in the normative and legal frameworks within which inter-state war exists. For any state actor involved in a conflict, this legal and normative context necessarily applies. As in dueling or in a wrestling match, the legal and normative context of war determines the limits of strategic actions that the actors in war can undertake. However, is there a similar context that adjudicates the behavior of opponents in the paradigm of war as a street brawl? If the argument that this study proposes is correct, it is likely that at least one side of any war is going to be populated with any number of volunteer foreign fighters, each of them participating in the war for reasons best known to only themselves. Can these foreign fighters be held accountable by the norms and formal laws of armed conflict? The behavior of strategic actors such as al Qaeda or the Islamic State already suggests that the formal laws of armed conflict are irrelevant to these strategic actors. This is a further complication, and potential strategic disadvantage, for the state actor facing a "flash mob."

If the argument that this study makes is correct, it means that wars in the 21st Century are likely to become even more complicated. A war may begin as a fairly straightforward *zweikampf*, but it may evolve into something rather more complicated, as otherwise uninvolved peoples from elsewhere become self-mobilized by what they read and what they see in print and on-line information sources. So what starts out resembling a duel or a wrestling match may very well degenerate into a street brawl.

The process that transforms a *zweikampf* into a street brawl is facilitated by the proliferation of information, and its consequent negative effect on cognition and understanding, allowing people to react in a visceral, even shallow and unthinking, manner to the images and information they are constantly bombarded with. Given that nationhood and national identity are constructed and imagined, it becomes increasingly easy for any person to identify with the struggles that another group is experiencing at that moment, the physical distances between them notwithstanding. One side may attempt to actively encourage such support from abroad. But even in cases where there is no such conscious attempt to generate overseas support, people from elsewhere may be self-mobilized based on what they read or see.

Strategy in such cases therefore becomes even more complicated than before. The purpose of strategy is to bring the opponent to the point where the opponent concedes defeat. In a *zweikampf*, such as the United States' war against Japan between 1941 and 1945, strategic success could be conferred upon the former when the latter lost its will to persevere. But the possession of otherwise overwhelming material resources by one side is not a sufficient condition to convince the opponent to concede defeat—witness the United States' war in Vietnam. To bring the opponent in a war framed as a duel to the point where he concedes defeat is already a difficult challenge for strategy. In a war framed as a street brawl, however, strategy has to address not one opponent, but multiple opponents, each coming to the war for a variety of reasons unique to each opponent; the common motive is likely the lowest common denominator. To bring these multiple opponents to the point of conceding defeat will surely multiply the challenges that strategy faces.

Endnotes

¹http://www.mitpressjournals.org/doi/pdf/10.1162/ISEC_a_00023.

²The Soufan Group, "Foreign Fighters: An Updated Assessment of the Flow of Foreign Fighters into Syria and Iraq," December 2015 https://thesoufancenter. org/wp-content/uploads/2017/05/TSG-Foreign-Fighters-in-Syria-Final-with-cover-rebrand-031317.pdf.

³Steven Metz, "The Price of Defeating the Islamic State," World Politics Review, 5 September 2014 http://www.worldpoliticsreview.com/articles/14036/the-price-of-defeating-the-islamic-state.

⁴Carl von Clausewitz (translated and edited by Bernard Brodie and Peter Paret), *On War* (Princeton, New Jersey: Princeton University Press, 1976), 75. In German, the first sentence of this quote reads, "Der Krieg ist nichts als ein erweiterter Zweikampf." http://www.clausewitz.com/readings/Compare/VomKriege1832/Book1Ch01VK.htm.

⁵http://www.sciencedaily.com/releases/2013/05/130522085217.htm; http://www. bbc.com/news/business-26383058.

⁶Lucas Mearian, "Worlds data will grow by 50X in next decade, IDC study predicts," *ComputerWorld*, 28 June 2011 http://www.computerworld.com/ article/2509588/data-center/world-s-data-will-grow-by-50x-in-next-decade--idc-study-predicts.html.

⁷https://www.statista.com/statistics/871513/worldwide-data-created/.

⁸https://www.weforum.org/agenda/2019/04/how-much-data-is-generated-each-day-cf4bddf29f/.

⁹http://www.internetworldstats.com/stats.htm.

¹⁰http://time.com/3604911/3-billion-internet-users/.

¹¹https://www.statista.com/statistics/617136/digital-population-worldwide/.

¹²http://www.statista.com/statistics/272014/global-social-networks-ranked-by-number-of-users/.

¹³Daniel J. Levitin, "Why the modern world is bad for your brain," *The Guardian*, 18 January 2015 http://www.theguardian.com/science/2015/jan/18/modern-world-bad-for-brain-daniel-j-levitin-organized-mind-information-overload. Also see: Jennifer Groff, "The Conflicted Brain: The Impact of Modern Technologies on Our Cognition, and How Arts Education can be the Keystone to Whole-mindedness," in E. Clapp (ed.), *20UNDER40: Reinventing the Arts and Arts Education for the 21st Century* (Bloomington: Author House, 2010), pp. 1–20.

¹⁴https://www.bbc.co.uk/blogs/digitalrevolution/2009/09/what-are-we-thinking-cognition.shtml.

¹⁵Kep-Kee Loh and Ryota Kanai, "How has the Internet reshaped human cognition?" http://www.researchgate.net/publication/280041294_How_Has_the_Internet_Reshaped_Human_Cognition.

¹⁶*Ibid.*

[17] Thomas Heghammer, "The Rise of Muslim Foreign Fighters: Islam and the Globalization of Jihad," *International Security*, vol. 35, no. 3, Winter 2010/2011, p. 69.

[18] Jack David Eller, *From Culture to Ethnicity to Conflict: An Anthropological Perspective on International Ethnic Conflict* (Ann Arbor: University of Michigan Press, 1999), pp. 16–17.

[19] Benedict Richard O'Gorman Anderson, *Imagined Communities: Reflections on the Origin and Spread of Nationalism* (London and New York: Verso, 2006), pp. 6–7.

[20] *Ibid.*, p. 135.

[21] Timothy J. Owens, Dawn T. Robinson, and Lynn Smith-Lovin, "Three Faces of Identity," *Annual Review of Sociology*, no. 36, 2010. http://www.annualreviews. org/doi/abs/10.1146/annurev.soc.34.040507.134725. Also see: Sheldon Stryker and Peter J. Burke, "The Past, Present, and Future of an Identity Theory," *Social Psychology Quarterly*, vol. 63, no. 4, December 2000, pp. 284–297.

[22] *Ibid.*, p. 490.

[23] See, for instance: Margaret R. Somers, "The Narrative Construction of Identity: A Relational and Network Approach," *Theory and Society*, vol. 23, no. 5, October 1994, pp. 605–649.

[24] *Ibid.*, p. 611.

[25] *Ibid.*, pp. 613–614.

[26] Owens, Robinson and Smith-Lovin, "Three Faces of Identity," p. 490.

[27] Bernd Simon and Bert Klandermans, "Politicized Collective Identity: A Social Psychological Analysis," *American Psychologist*, vol. 56, no. 4, April 2000. p. 321. http://dare.ubvu.vu.nl/bitstream/handle/1871/33665/139773.pdf?sequence=1.

[28] Anderson, *Imagined Communities*, p. 7.

[29] Mary Kaldor, "Elaborating The 'New War' Thesis," in Isabelle Duyvestyn and Jan Angstrom (ed.), *Rethinking the nature of War* (London and New York: Frank Cass, 2005), p. 211.

[30] H. Tajlel (ed.), *Differentiation between social groups* (London: Academic Press, 1978), p. 86.

[31] Simon and Klandermans, "Politicized Collective Identity," p. 20.

[32] Joel Kotkin, *Tribes: How Race, Religion, and Identity Determine Success in the New Global Economy* (New York: Random House, 1992), p. 16.

[33] *Ibid.*, p. 322.

[34] John C. Turner and Katherine J. Reynolds, "Self-Categorization Theory," in Paul A. M. van Lange, Arie W. Kruglanski and E. Tory Higgins (eds.), *Handbook of Theories of Social Psychology 2* (London and Thousand Oaks: Sage Publications, 2012), pp. 399–417.

[35] Daniel Milton, *Communication Breakdown: Unraveling the Islamic State's Media Efforts* (West Point: United States Military Academy, 2016), p. 5.

36 Thomas Rid and Marc Hecker, *War 2.0: Irregular Warfare in the Information Age* (Westport, Connecticut and London: Praeger Security International, 2009), p. 8.

37 Simon and Klandermans, "Politicized Collective Identity," p. 324.

38 Heghammer, "The Rise of Muslim Foreign Fighters," pp. 57–58.

39 *Ibid.*, 62.

40 David Malet, "Why Foreign Fighters: Historical Perspectives and Solutions," *Orbis Journal of Foreign Affairs,* vol. 54, no. 1, Winter 2010. p. 101.

41 Cary Nelson and Jefferson Hendricks (eds.), *Madrid 1937: Letters of the Abraham Lincoln Brigade From the Spanish Civil War* (London and New York: Routledge, 1996), p. 2.

42 *Ibid.*, p. 29.

43 *Ibid.*, pp. 29–46.

44 Peter N Carroll, *The Odyssey of the Abraham Lincoln Brigade: Americans in the Spanish Civil War* (Stanford: Stanford University Press, 1994), pp. 4–5.

45 R. Dan Richardson, *The International Brigades and the Spanish Civil War* (Lexington: University Press of Kentucky, 1982), p. 1.

46 *Ibid.*, p. 7.

47 Heghammer, "The Rise of Muslim Foreign Fighters," pp. 83–84.

48 *Ibid.*, 67.

49 One such warlord was the so-called Lion of Panjshir, Ahmad Shah Massoud, who eventually became Minster for Defence between 1992 and 1993. See: Steve Coll, *Ghost Wars: The Secret History of the CIA, Afghanistan, and bin Laden, from the Soviet Invasion to September 10, 2001* (London: Penguin Books, 2005), p. 10.

50 Bureau of Intelligence and Research, US Department of State, "The Wandering Mujahidin: Armed and Dangerous," 21–22 August 1993. http://blogs.harvard.edu/mesh/files/2008/03/wandering_mujahidin.pdf.

51 Joseph A. Carter, Shiraz Maher, and Peter R. Neumann, "#Greenbirds: Measuring the Importance and Influence in Syrian Foreign Fighter Networks," The International Centre for the Study of Radicalisation and Political Violence, King's College, London, 2014, p. 2. http://citeseerx.ist.psu.edu/viewdoc/download; jsessionid=14EF9111051335DA9548FF4EB8C89E16?doi=10.1.1.662.1555&rep=rep1&type=pdf.

52 Adam Rawnsley, "Meet the Americans Flocking to Iraq and Syria to Fight the Islamic State, *Foreign Policy,* 26 August 2015 https://foreignpolicy.com/2015/08/26/meet-the-americans-flocking-to-iraq-and-syria-to-fight-the-islamic-state/.

53 Rid and Hecker, *War 2.0*, p. 129.

54 *Ibid.*, pp. 125–206.

55 *Ibid.*, pp. 134–136.

56 *Ibid.*, p. 100.

[57] Malet, "Why Foreign Fighters?," p. 112.

[58] "Combatants on Foreign Soil," Second International Conference on DDR and Stability in Africa, Kinshasa, Democratic Republic of Congo, 12–14 June 2004 https://au.int/sites/default/files/documents/39195-doc-150._second_international_conference_on_drr_in_africa_kinshasa_drc._combatants_on_foreign_soil.pdf.

[59] Bernard Fook Weng Loo, "Decisive Battle, Victory, and the Revolution in Military Affairs," *Journal of Strategic Studies*, vol. 32, no. 2, March 2009, pp. 189–211.

[60] See, for instance, Judith A Nicholson, "Flash! Mobs in the Age of Mobile Connectivity," in *The Fibreculture Journal*, no. 6, 2005 http://six.fibreculturejournal.org/fcj-030-flash-mobs-in-the-age-of-mobile-connectivity/.

[61] Bill Wasik, "My Crowd: Or, Phase 5: A report from the inventor of the flash mob," *Harper's Magazine,* March 2006, p. 57.

[62] See: http://news.yahoo.com/islamic-state-recruits-broadly-not-just-fighters-081502457--politics.html; http://www.pbs.org/newshour/rundown/islamic-state-lures-westerners-humanitarian-approach/.

[63] George Lakoff, "The Contemporary Theory of Metaphor," in Andrew Ortony (ed.), *Metaphor and Thought* 2nd ed. (Cambridge: Cambridge University Press, 1993), p. 203 http://www.cogsci.ucsd.edu/~coulson/203/lakoff_ps.pdf. Also see Michael Reddy, "The Conduit Metaphor-A case of Frame Conflict in Our Language about Language," in Andrew Ortony (ed.), *Metaphor and Thought* (Cambridge: Cambridge University Press, 1979), pp. 284–310.

[64] George Lakoff, "Metaphor and War, Again," 17 March 2003 http://www.alternet.org/story/15414/metaphor_and_war,_again. Also see: Roland Paris, "Kosovo and the Metaphor War," *Political Science Quarterly*, vol. 117, no. 3, Fall 2002, p. 427.

[65] Cited in Paris, "Kosovo and the Metaphor War," p. 427.

[66] Rupert Smith, *The Utility of Force* (London: Allen Lane, 2005), p. 3.

[67] Paris, "Kosovo and the Metaphor War," p. 425.

[68] *Ibid.*, p. 413.

[69] Yuen Foon Khong, *Analogies at War: Korea, Munich, Dien Bien Phu, and the Vietnam Decisions of 1965* (Princeton, New Jersey: Princeton University Press, 1992).

[70] Clausewitz, *On War*, pp. 75, 77.

[71] George Lakoff, "Why It Matters How We Frame the Environment," *Environmental Communication*, vol. 4, no. 1, March 2010, p. 72. Lakoff argues, "since frames come in systems, a single word typically activates not only its defining frame, but also much of the system its defining frame is in."

[72] Bernard F.W. Loo, "Decisive Battle, Victory, and the Revolution in Military Affairs," *Journal of Strategic Studies*, vol. 32, no. 2, April 2009, pp. 189–211.

[73] See, in particular: Alan Beyerchen, "Clausewitz, Nonlinearity, and the Unpredictability of War," *International Security*, vol. 17, no. 3, Winter 1992/1993,

p. 73. Also see: Peter Watson, *War on the Mind. The Military Uses and Abuses of Psychology* (Harmondsworth: Penguin, 1980); Frank Goldstein and Benjamin F. Findley, Jr. (eds.), *Psychological Operations: Principles and Case Studies* (Maxwell AFB: Air University Press, 1996); Randall G. Bowdish, "Information-Age Psychological Operations," *Military Review*, vol. 78, no. 6, Dec. 1998/Jan-Feb 1999, pp. 29–37.

[74] David Kilcullen, *The Accidental Guerrilla: Fighting Small Wars in the Midst of a Big One* (Oxford and New York: Oxford University Press, 2009), p. 10.

[75] *Ibid.*, p. 34.

[76] *Ibid.*, p. 263.

[77] Rid and Hecker, *War 2.0*, pp. 128–129.

[78] Frank Hoffman, "Hybrid Warfare and Challenges," *Joint Forces Quarterly*, no. 52, 1ˢᵗ quarter 2009, p. 35.

[79] *Ibid.*, p. 37.

[80] *Ibid.*, p. 38.

[81] Department of Defense, *Quadrennial Defense Review 2014*, pp. 19–22.

[82] *Quadrennial Defense Review 2014*, p. 3.

[83] The National Defense Panel's review of the 2014 QDR notes that technological diffusion "will likely continue to allow non-state actors and even individuals to prosecute more aggressive terrorist and criminal operations." In the estimation of the Panel, therefore, individuals matter strategically only in the context of terrorist and criminal actions, not as volunteers and foreign fighters who can transform a two-sided war. See: William J Perry *et al.*, *Ensuring a Strong U.S. Defense for the Future: The National Defense Panel Review of the 2014 Quadrennial Defense Review* (Washington, D.C.: United States Institute of Peace, 2014), p. 21.

[84] John Mackinlay, "International Operations to Contain Violence in a Complex Emergency," in Isabelle Duyvestyn and Jan Angstrom (eds.), *Rethinking the nature of War* (London and New York: Frank Cass, 2005), p. 179.

[85] See, for instance: Stephen Metz, "U.S. Military Must Transcend Warfighting Mindset," *World Politics Review*, 5 December 2012 http://www.worldpolitics review.com/articles/12543/strategic-horizons-u-s-military-must-transcend-warfighting-mindset; David Vergun, "Warfighting capacity at increasingly worrisome levels, says Army planner," 25 November 2014 http://www.army.mil/ article/138813/Warfighting_capacity_at_increasingly_worrisome_levels__says_ Army_planner/; "Troops get back to basics on Salisbury Plain" https://www.gov. uk/government/news/troops-get-back-to-basics-on-salisbury-plain.

[86] Clausewitz, *On War*, p. 593.

[87] John Drury, "Crowd Psychology," in T. Teo (ed.), *Encyclopedia of critical psychology* (New York: Springer, 2014), pp. 341–344 https://www.academia. edu/931498/Crowd_psychology.

[88] Stephen Reicher, "The Psychology of Crowd Dynamics" http://www.uni-kiel. de/psychologie/ispp/doc_upload/Reicher_crowd%20dynamics.pdf.

[89] *Ibid.*

[90] *Ibid.*

[91] David Schweingruber, "Mob Sociology and Escalated Force," *The Sociological Quarterly*, vol. 41, no. 3, 2000, p. 374 http://davidschweingruber.com/docs/ Schweingruber(2000)-TSQ.pdf.

[92] Reicher, "The Psychology of Crowd Dynamics," p. 6.

[93] Homi K. Bhabha, "Narrating the Nation," in Vincent P. Pecora (ed.), *Nations and Identities* (Malden and London: Blackwell Publishers, 2001), p. 359.

[94] Michael Ignatieff, *Blood and Belonging: Journeys into the New Nationalism* (New York: Farrar, Straus and Giroux, 1993), p. 5.

[95] Rid and Hecker, *War 2.0*, p. 9.

[96] Barry D. Watts, *Clausewitzian Friction in Future War* (Maxwell, Alabama: Air University Press, 1996).

Chapter 6

Military Engagement in Disaster Response Policies, Interests and Issues

Alistair D.B. Cook

Senior Fellow, Centre for Non-Traditional Security Studies,
S Rajaratnam School of Internal Studies,
Nanyang Technological University, Singapore

iscook@ntu.edu.sg

Over the past twenty years, the world has witnessed large-scale disasters which have initiated significant international humanitarian efforts to provide disaster relief to affected countries. In 2003 the Indian Ocean Earthquake and Tsunami caused widespread disruption from the east coast of Africa to the Asia-Pacific. It saw the activation of militaries, civilian and UN agencies, NGOs and the Red Cross and Red Crescent Movement to assist affected governments and people. Five years later, Myanmar was affected by Cyclone Nargis in 2008 which left 138,366 people dead and 1.5 million in the Irrawaddy delta severely affected. This time, the international humanitarian effort was initially stymied by the reluctance of the military junta in power in Myanmar to admit foreign military forces into the country, and was made possible only by the personal diplomacy of Southeast Asian officials through ASEAN under the Tripartite Core Group which facilitated the humanitarian response from the international community to those affected in Myanmar.

In 2010, the Haiti Earthquake devastated the country with an estimated 200,000 dead and 3 million affected. In 2011, Japan was struck by a triple disaster of an earthquake, tsunami and a meltdown at its nuclear power plant. This disaster saw an estimated 156,000 people displaced and the Japanese government overwhelmed by the scale of devastation as well as institutional challenges to engage the international community's humanitarian support. In 2013, Super Typhoon Haiyan killed 6,340 people and severely affected about 11 million people. It devastated parts of the Philippines and the widespread level of destruction initiated an international response drawing on diverse actors from civilian agencies, militaries, NGOs, private sector and the Red Cross and Red Crescent Movement. In 2016, Tropical Cyclone Winston hit Fiji the hardest and was the strongest cyclone on record in the Pacific displacing 350,000 people approximately 40% of its population. All these disasters initiated significant international responses involving militaries. Why are the military involved in disaster response? What are the parameters for such an engagement? What are the most recent developments in international disaster response to illustrate current dynamics in the field?

This chapter will proceed to answer these questions in the following ways. First it is important to understand the disaster context, what defines a 'disaster', is it contested, and in what ways do we understand the term impacts military decision-makers? Second, this chapter will survey the major analytical frameworks that inform scholarly debates and capture the motivations behind military decision-making in relation to disasters. Third, the chapter will assess the international context in which these disasters and military decision-making occur to identify the dominant global frameworks shaping international relations on disaster response. Fourth, the chapter will highlight these global dynamics through case studies to illustrate military engagement in disaster response and the challenges faced by disaster responders. Finally, this chapter will conclude with some of the emerging directions and challenges faced by military decision-makers determining involvement in disaster response.

Conceptualizing Humanitarian Assistance and Disaster Relief (HADR)

What explains military involvement in HADR? Up until the 1970s disasters were primarily understood as solely physical occurrences requiring technological solutions. Since then policy makers and scholars of disaster studies have understood them, and the solutions needed, through the lens of vulnerability. Vulnerability is an individual's, a household's, a community's, or a society's exposure to a natural hazard (a geological event like earthquakes or a meteorological event like typhoons and cyclones). This approach identifies the way human systems place people at risk in relation to each other and the environment.[1] As a result, this shift refutes the claim that disasters are natural as the occurrence of a natural hazard will expose certain communities within countries more than others as a result of their place in society. It therefore calls for situational awareness of the pre-existing social dynamics of an area affected by a natural hazard to inform an international humanitarian response effort.

The occurrence of disasters has activated national civilian government agencies mandated to oversee issues like development, agriculture, civil defense and disaster management. In some instances, the military has been activated to support these national efforts when the civilian agencies need support or are incapacitated directly by the disaster or by the acute needs of the affected people in the aftermath of a disaster. In other instances, common in Asia, the militaries are deployed as a first responder. This contrasts with the human consequences of conflict which have primarily seen governments as party to a conflict and the need for a response outside of the conflict setting, which traditionally activates humanitarian agencies and other actors perceived as neutral, independent and impartial working to fulfil the needs of communities affected by, but not directly engaged in, a conflict.

The understanding of disaster as an event that affects people dependent on their place in society underlines the human cause of the disaster itself. In the case of ASEAN, the regional organization agreed in 2008 the ASEAN Agreement on Disaster Management and Emergency Response (AADMER). In this agreement, disaster means "a serious disruption of the functioning of a community or a society causing widespread human, material, economic or environmental losses." Likewise, a hazard refers to "a potentially damaging physical event, phenomenon and/or human activity, which may cause the loss of life or injury, property damage, social or economic disruption or environmental degradation."[2] This agreement has been invoked for both natural hazards and conflict settings as seen in the annual effects of cyclones, typhoons, earthquakes and volcanic eruptions. AADMER was also activated to govern the response effort in the internal conflict cases of the Siege of Marawi in the Southern Philippines and the Rakhine Crisis in Myanmar in 2017 illustrating its broader applicability beyond solely natural hazards.

These cases not only illustrate the overlaps between disaster and conflict settings in the provision of humanitarian assistance and disaster relief, but also importantly signal the emerging lead role given to disaster management agencies in both cases as a component part of the national government in ASEAN Member States. These examples underline the importance of the framing of a response effort in understanding the civilian and military actors involved and the roles they play in relation to the context in which they provide humanitarian assistance and disaster relief to affected communities. This is not a foregone conclusion but rather highlights the inaction of the relevant peace pillar within ASEAN to address conflict situations and provides a stop-gap solution.

Further the experience of individual countries and regions to either conflict or disaster pre-ordains the discussion of the military. In the Asia-Pacific which is home to around 80% of the world's disasters, understandably much focus has been on the development of international, regional, national, and local responses to natural hazards over developing mechanisms for humanitarian assistance in conflict settings. This compares to the global humanitarian system where 80% of response efforts are in conflict zones.

This is further complicated when national, regional or international systems tailored to the dominant natural hazard experience are imposed upon internal conflict settings. In Southeast Asia, the regional organization has a well-developed regional (civilian) framework that is supported by militaries in times of disaster as requested. However, the AADMER definition of disaster has seen the framework activated in conflict settings notably the Rohingya Crisis since 2017 and the Siege of Marawi also in 2017. In both contexts the regional framework was used to coordinate the provision and delivery of humanitarian assistance and was handed over to government officials. This was necessary as the regional framework is governed by ASEAN Member States and is essentially a state-based humanitarian framework within a multi-stakeholder environment.

Herein lies a definitional problem—what constitutes a 'disaster' now generates a significant debate as the disaster management and emergency response framework is activated by conflict as understood as a 'human-induced' disaster. This debate largely occurs within an environment where there is an absence of political buy-in to govern conflict settings, as seen in Southeast Asia.

It is important to note that the International Red Cross and Red Crescent Movement has long had parallel but connected structures to govern their involvement in conflicts and disasters. The International Committee of the Red Cross (ICRC) holds the mandate and framework for engagement in conflict settings, whereas the International Federation of the Red Cross and Red Crescent (IFRC) holds the mandate and framework for engagement in disaster settings. When there are clear differences between the two settings, involvement is less contentious and supporting a national government after a natural hazard is the default whereas the humanitarian principles of humanity, neutrality, impartiality and operational independence are the cornerstone of humanitarian action in conflict settings to avoid acting on behalf of, or being seen to act on behalf of, one or other of the parties to a conflict.

For the purposes of this chapter, the focus is on natural hazards, but with the caveat that decision-makers need to have situational awareness of the underlying causes of the disaster so they can identify challenges and constraints in advance. This will require a longer-term investment in monitoring and evaluation of the security landscape and its implications for humanitarian assistance and disaster relief.

Pay-offs and Trade Offs: An Analytical Framework

The involvement of the military in disaster response has become commonplace since the turn of the 21st century. Indeed, the battle to keep the military out of disaster response was lost long ago, and the conversation has moved on to civil-military complementarity and niche expertise.[3] There are several motivations for military involvement in disaster response from its surge capacity allowing for quick response efforts, to fulfilling bilateral and multilateral commitments to security partners and allies, providing training for service personnel in a field environment, to showcasing military capacity and force projection capabilities. This section outlines ways in which to understand the motivations of countries involving their militaries in disaster response.

Malesic analyzed the theoretical considerations of military deployment in disaster response, and categorized five core arguments most frequently invoked in the material under investigation as utilitarian, security-strategic, militarization, functional-humanitarian, and rejection-isolation.[4] The militarization and utilitarian arguments are the most popular in the analysis. Militarization warns that "an increased military presence in disaster management may threaten the operating principles of civilian agencies and humanitarian organizations and may lead to their replacement with military structures."[5] However, the utilitarian approach underplays the role institutional structures, rules, norms and cultures play as part of the position the affected country's military has in the national experience. Scholars like Muthiah Alagappa[6] identified the early stage of state formation in Southeast Asia with the legitimacy of national governments being challenged by internal armed groups and the crucial role the military played in state and nation building and regime stability. The militaries have taken on developmental roles involved in local communities and so have developed significant, positive or negative, relationships dependent on context.[7] This is particularly important when investigating the role of militaries in disaster response as both sender and recipient forces and so drawing on the history and place of militaries within national experiences, particularly those outside the Global North, are instructive in this regard particularly as national capacities and institutions further develop.

While these categories offer an understanding of the dominant approaches within the civil-military literature, they ignore two important components to understanding the role of militaries in disaster response efforts particularly in the Asia-Pacific. While the utilitarian approach emphasizes the benefits to the deploying country by engaging its troops on disaster sites in a foreign country, it underplays the role of the military within the affected country. It does recognize however the positive impact that military engagement has on the military image and legitimacy in the public realm of the affected country.[8] The rejection-isolation argument can be grouped as rejectionist from within the military to 'mission creep' of the armed forces in areas outside of its remit. The utilitarian argument has been referred to as 'disaster militarism' as it advocates militaries looking beyond their core mandate as department budgets and austerity measures limit the amount of investment in the core mandate of militaries and so engaging in non-traditional missions can make up the shortfall.[9]

Further, critical scholarship plays an important role in understanding the structural context of military involvement in disasters. Feminist peace scholars, for example, argue that militaries legitimize vast amounts of money spent on deadly weapons; while limiting the public imagination of bravery and service to military (usually male) personnel. These ideas about soldiering and the military distort our understandings of community, safety and security where greater recognition of contributions and perspectives from across society is underappreciated. Instead of gifting militaries with extra resources during crises, we should invest in national and transnational, non-militarized and non-masculinized, appropriately trained, civilian corps to do the job.[10] This provides decision-makers with an important consideration as to whether civilian capacity is better positioned now and, in the future, to provide disaster response. However, the answer to this can cut both ways.

In Indonesia after the performance of the national disaster management agency (BNPB) in leading the response to several disasters, President Joko Widodo removed the restriction on serving military officers being appointed to BNPB. This is often characterized as the militarization of the civilian national disaster management agency under the aegis that the military officers are better equipped to lead national disaster response efforts. This shift within the civilian administration is opposed by some civilian bureaucrats who cite the issue as more to do with its mandate than leadership. As a national civilian agency, it was established to perform a firefighting role to send in civilian support to disaster affected areas. However, unlike most national agencies and ministries it does not have a sub-national support structure and therefore has weak links on the ground. Its role in several disaster responses was focused on the financial management of incoming overseas assistance. The proponents of the rejection-isolation argument argue that disaster response is a distraction from the core function of the military and reduces its capacity for combat missions or so-called 'mission creep'.[11] This often comes from within the military ranks who see the professionalization of the military to be combat ready and need niche combat capabilities and specialization rather than general exposure to field missions. So the 'rejection-isolation' proponents argue that militaries should focus on their core responsibilities as combatants in war. However, Djuyandi, Casnoto, and Hidayat argued that there was a lack of synergy that had evolved between BNPB and the TNI in training and that greater synergy needs to be further developed.[12] In the case of Indonesia there are those who argue for the development of civilian capabilities to disaster management and those that acknowledge the benefits that military involvement can bring to a disaster response effort.

The utilitarian argument as identified by Malesic does not advocate the use of military in disaster response based on the needs of the affected population, but rather on the benefits accruing to the states and militaries that offer such assistance, i.e., foreign militaries.[13] In this sense the deployment of militaries overseas could offer benefits to the sending military in the form of field experience where soldiers who would otherwise be at home base, participating in exercises and simulations could instead engage in a field mission which would offer greater insight into military organization and interacting with local populations. However, the argument assumes a willing recipient of external military support for an affected country's domestic disaster. Ultimately disasters know no nation-state boundaries, but their effects do fall within one or more of them. As a result, the recipient government will necessarily approve external military support and so the successful deployment of militaries in disaster response is contingent on the relationship to the host government and military. David Capie nuances the instrumental benefits of HADR to reflect this by grouping the benefits into three categories: signaling commitment and helping facilitate access, demonstrating capabilities, and providing opportunities for coalition building and engagement with new partners.[14] While this would better prepare the soldiers to be better combat ready for future mission deployment for a military sending country, the benefits to the affected people remain unaccounted for in this calculation.

The less frequently invoked arguments according to Malesic are the security-strategic, functional-humanitarian, and rejection-isolation arguments. The security-strategic argument holds that international military assistance in disaster response tends to promote better relations with allies and to strengthen the position of the country in the regional security architecture.[15] This argument is often present when major powers engage affected countries with whom they have longstanding relations often in the form of Visiting Forces Agreements or Status of Forces Agreements and trusting relations that facilitate the rapid mobilization of a HADR effort. This cannot be discounted when considering that the first 72 hours after a major disaster are critical in response time, after which those affected if left without assistance will have significantly reduced chances of survival.

The development of these security partnerships and alliances can prepare the necessary legal channels to facilitate HADR engagement by militaries.[16] One example is the ANZUS states of Australia, New Zealand and the United States cooperating in the provision of HADR in the Pacific.[17] Further, the notion of issue-linkage—in this case military engagement in disaster response can improve strategic-security ties with recipient countries—is often identified as justification or an additional benefit. The United States military engagement in the disaster response to Super Typhoon Haiyan in the Philippines for example was used to justify the increased US military presence in the Philippines and added weight to the need for a Mutual Defence Agreement.[18]

While each of these arguments offers a partial truth, they collectively offer important facets into the motivations for military involvement in HADR operations. The role of international norms suggests that military engagement in disaster response is accepted practice although the contributions remains negotiated depending on individual militaries niche capabilities. Within the international legal system, developments are underway to formalize the military role in disaster response such as through International Disaster Relief Law. However, to understand why countries sometimes participate and other times do not participate, we need to look at both the individual country's domestic context and its place within the international system. However, international relations and strategic studies literature primarily focuses on the deployment of militaries overseas. This continuing focus underlines one of the significant challenges of the humanitarian system which is that it remains a supply-driven rather than a needs-based model.[19]

As power shifts within the international system with the economic and military development of individual countries, their expanding and contracting alliances, and their force and image projection, will offer a more comprehensive understanding of the trade-offs that countries make in determining their contributions to disaster response operations and the reception to such offers by disaster-affected states. While military capacity goes some way to explain their niche capabilities with dual-use or hybrid assets that can be deployed to disaster response operations, alone this is not enough. There needs to be a holistic understanding of the motivations and trade-offs that states make in the contributions to disaster response operations.

The functional-humanitarian argument holds that civilian capacities can be overwhelmed as a result of a disaster and as such national and foreign militaries are well-placed to fill this gap. This argument is often heard within civil-military coordination units within the United Nations, or variants of it, where militaries with a longer history of HADR engagement, UN agencies and NGOs with experience in major disaster settings have witnessed the positive aspects of humanitarian civil-military cooperation within an international framework that seeks to draw on multistakeholder niche capabilities.

HADR missions do not tend to expose armed forces personnel to combat, but nonetheless they experience many war-like situations such as

death, disease, destruction of homes and cities, political and social insecurity, and suffering populations. While mission types differ substan-
tively, the effects on individual service personnel are broadly similar
compared to those who do not deploy but the hazard of PTSD and death
by suicide are quite different between combat and HADR missions.[20] This
argument is often articulated to justify military involvement in disaster
response as beneficial to its core mandate to develop combatants.

However, these additional benefits to a military's core mandate are
not alone. Canyon, Ryan and Burkle argue that the role of the military in
HADR goes beyond the emerging partnership on logistics to more broadly
encompass the disaster management cycle and support the Sendai
Framework for Disaster Risk Reduction (2015–2030).[21] These authors
envisage positive contributions from militaries in prevention activities
such as dam construction to avert flooding, and recovery phase to restore
livelihoods; public health infrastructure; and economic, social, and environmental systems as part of the commitment to "Build Back Better."[22]
Hornyacsek found that militaries already have the capabilities to engage
in disaster management in Europe.[23] In the Asian context, this is recognizable in the state-building and development efforts that militaries have
engaged in the post-colonial period. However, it is also observable that
military experience in overseas disaster response also contributes to their
greater leadership role in domestic disaster response.[24] However, the
military role in disaster response receives push back from civil society
which calls for more investment and decision-making power in civilian
governance structures.

Beyond the military's suitability to disaster response and broader disaster management activities, some argue that the use of the military makes
good use of available assets. The utilitarian argument is often focused on
operations management and product or system 'utilization' in disaster settings using econometrics to quantify variables including asset capabilities,
proximity, and the duration of an HADR response to find a 'most effective' response evaluation particularly when there are two similar assets
available with similar effects but significantly different costs.[25] Arguably
therefore militaries purchase more dual-use conventional military capabilities to justify expenditure. Japan, South Korea and China for example
now possess the willingness to involve their militaries in international
non-traditional security roles, but they are also acquiring significant dual-
use conventional military capabilities such as heavy airlift and large
amphibious vessels to make a substantial contribution.[26] However, this

occurs when financial tracking of assets and personnel deployment is captured which is not systematically carried out. Indeed, the United States Department of Defence does not set aside funds specifically for disasters. The obligation of such funds is disaster dependent.

However, the United States Navy now includes Humanitarian Assistance and Disaster Relief as a core capability as described in *A Cooperative Strategy for 21st Century Seapower* and a key national strategic priority in the 2012 Defense Strategic Guidance, 2014 Quadrennial Defense Review (QDR), and 2015 National Security Strategy.[27] While incremental costs (additional food, fuel and temporary additional service personnel) to the DoD budget are insignificant, there are significant 'submerged' costs (recruiting and training extra troops, purchasing and servicing additional equipment, additional layers of bureaucracy, and maintaining and enlarging bases) which are not included in the report to the US Congress.[28] The report also does not include the costs of service personnel who would be on a vessel when a disaster occurred and the availability of the vessel itself.

However more recent calls have been made for greater transparency over the costs involved in utilizing military assets for HADR now that internal markets in government mean that budgets of non-military ministries are billed for the deployment of military assets in HADR scenarios. This becomes more contentious when the deployment of civilian aircraft or transportation using private sector alternatives offer better value-for-money. As Apte and Yoho identify, vertical lift as the most expensive component of the United States Navy contributions to HADR.[29] Ballesteros, Useem and Wry argue that there is a significant role for private sector organizations, local companies providing relief in a disaster affected area, are quicker at recognizing needs and reconfiguring their resources to meet those needs, whereas 'traditional aid providers' are slower off the mark but are still necessary.[30] Yuste *et al.* highlight the complementarity of humanitarian, private sector and military contributions to HADR efforts and signal a trilateral synchronization of efforts as the next step in disaster response.[31] Indeed, this is visible in the private sector participation in the Regional Consultative Group on Humanitarian Civil-Military Coordination in the Asia-Pacific, most notably the active participation of the DHL Disaster Response Team. This raises questions over (1) what is the full financial cost of military engagement? (2) If there are more cost-effective civilian alternatives such as commercial flights available then what are the other considerations that justify military engagement? Hofmann and Hudson observe that military involvement in disaster management is driven by the military's own concern to improve its public image, to use disaster response as a form of training and to diversify the military role in times of austerity and military budget cuts.[32]

While civilian agencies and the private sector play an increasingly important role in disaster response offering an alternative to military assets, there are other considerations in the decision-making process of engaging militaries. In Canada's response to the Haiti Earthquake in 2010, the Canadian Government deployed C-17s. The government had earlier received criticism over the cost of these assets and saw their deployment as an important opportunity to show-case their benefit. A Department of National Defence presentation indicates that the Chief of Defence Staff (CDS) of the CF gave the order to "Go Fast, Go Big."[33] This highlights the role that public perception of the military can provide justification to decision-makers for military involvement in disaster response. This is a common feature across the realm of humanitarian work often regarded as the 'CNN Effect' whereby states and NGOs want visibility for their organizations as it is tied to presence in the case of states or to their ability to fundraise with the viewing public on the part of NGOs. This remains as true today as it was in the immediate aftermath of the Cold War.

As part of falling support for Canada's involvement in Afghanistan, the significant response to the Haiti Earthquake in 2010 saw news coverage divert public attention away from its controversial involvement in Afghanistan to the Haiti Earthquake response which was an ideal way to highlight Canada's global leadership and potentially rebuild public trust in the government.[34] Further when governments provide direct financial assistance to an affected country there is no "national stamp" to show their presence. However, with the use of military assets and personnel the national flag is clearly displayed offering a clear visual representation of a donor government's activity. It can further impact public opinion that military response is the best—or only—way to respond to disasters.[35] Now social media presence through dominant mediums like Facebook and Twitter provide a new avenue to engage the public, potential individual donors to activate civilian capacity, and raise awareness of issues to encourage greater political will to address challenges faced by vulnerable populations. As a result, there are several considerations decision-makers face when evaluating the deployment of militaries in disaster response.

Evolution of Global Humanitarian Operational Environment

In a disaster response there are often a range of actors involved from militaries, civilian agencies, NGOs, Red Cross and Red Crescent Movement, faith-based groups, civil society organizations, and the private sector to name the most salient. However, the term 'humanitarian' has a long history that dates to the Battle of Solferino in 1859 which led Henri Dunant to begin the process that led to the Geneva Conventions and founding of the Red Cross. The core humanitarian principles on which these engagements are based are, namely (1) humanity—to treat all humanely and equally by saving lives and alleviating suffering. More recently this was recrafted as 'the humanitarian imperative' to include people affected by natural hazards. (2) Impartiality—providing aid regardless of someone's social grouping (nationality, race, religion, gender, political opinion or class). (3) Neutrality—to not take sides in any hostilities or engage in controversial activities of an ideological, political, religious or racial nature. (4) Independence—to have operational independence and form their own policies and procedures independent of government. This view of humanitarianism is founded on the logic that humanitarian assistance is a universal right, victims are those without assistance, and humanitarian agencies assert their right to access the affected population. This often means it is done on their own terms with little reference to the policies, concerns and capacities of the affected state.[36]

This is in stark contrast to the mandate and workings of militaries which are the traditional parties to interstate conflicts. In the post-Cold War period, it is within this context that third party militaries offered support to international humanitarian efforts. The contentious involvement of the military in disaster response has been so because international law treated the military involvement in humanitarian affairs comprehensively including both conflict and disaster with only disparate reference to their involvement in disaster response alone. As a result, the debate of military involvement has often centered on their partial nature in providing humanitarian aid because in a war zone a military actor would be perceived as allied to one side or the other. The impact of the international legal system on the involvement of the military in humanitarian assistance in war zones has shaped the way in which its involvement is perceived as negative in disaster zones. The issue is further complicated when pre-existing tensions exist in areas affected by disaster. In such 'complex'

humanitarian emergencies the use of the military is much more contentious. As a result, there are several different frameworks in operation to
govern disaster response which are often complementary but can also be
in competition with one another. Taking international laws applying to
conflict zones governing humanitarian assistance and applying them to
disaster zones need to be distinct and relevant. To place both scenarios
under the same umbrella is "foolhardy, short-sighted and incompetent."[37]

In 2007 the IFRC developed the "Guidelines for the Domestic
Facilitation and Regulation of International Disaster Relief and Initial
Recovery Assistance" (IDRL) that brought together different components
of international law into one document to provide guidance on the legal
foundations of disaster response. Over the past decade the International
Law Commission has further formalized this effort with the development
of the 'Protection of Persons in the Event of Disasters' as a legally binding
framework scheduled for debate and approval by the United Nations
General Assembly in 2020–2021. The Geneva Conventions are the basis
for International Humanitarian Law or the Laws of War yet the spirit of
this body of law currently forms the reference point for disasters as well.
However, there have been a long series of important documents that have
been developed to govern international disaster response, resulting in
the IFRC International Disaster Response Law, Rules and Principles
Programme (IDRL) Guidelines in 2007. It has subsequently taken until
2020 for the 'Protection of Persons in the Event of Disasters' to be depos-
ited at the United Nations General Assembly for agreement. While IDRL
is more a synthesis of laws, rules and principles already in effect, there
presently has not been a reference document agreed at the United Nations
to govern disaster response until now.

First, military actors are becoming increasingly keen on engaging in
disaster response overseas. Secondly, military responders face many of the
same obstacles of civilian responders in disaster relief operations. As noted
in one summary of the "lessons learned" from NATO's intervention in the
2005 Pakistan Earthquake. "[the importance of working with host governments must not be underestimated. Many issues must be resolved before
operations forces arrive, including terms of entry, force protection, legal
status. communication channels. liaison arrangements, contracting arrangements. use of land for basing and translators.[38] Third, militaries face a similarly normative patchwork as found by civilian agencies.[39] Thus
contextualization of the affected area by disaster is important and this
requires pro-active engagement and knowledge-building of areas at risk of

disaster to ensure adequate situational awareness. This is despite the history of military involvement in disasters and the absence of warring parties in many disasters. Military engagement in humanitarian action in conflict settings was often legitimized through a UN Security Council resolution throughout the 1990s but much less often today. In disaster settings, there is more prominence to national government calls for assistance although the UN provides its own monitoring and evaluation of disasters.

Civil-Military Coordination

In 1992 the UN General Assembly passed Resolution 46/192 which established the Office for the Coordination of Humanitarian Affairs (OCHA) to bring together different components of UN specialized agencies and NGOs and provide a platform for Civil-Military Coordination. Alongside the establishment of UN OCHA, the Steering Committee for Humanitarian Response, a coalition of international humanitarian NGOs (established in 1972) launched the Code of Conduct for International Red Cross and Red Crescent Movement and NGOs in Disaster Relief in 1992. This body shares its members experience with UN specialized agencies through the UN's Inter-Agency Standing Committee (IASC). This led to the establishment of the Sphere Standards in 1997, initially technical guidance on expected standards of equipment in all humanitarian settings. It has periodically revised its handbook with the latest iteration launched in 2018 focused more on context analysis and empowering local actors in their own recovery.

In the early 2000s, the post-Cold War UN-centric system was challenged in the aftermath of the September 11th 2001 terrorist attacks on the United States and the subsequent US-led invasions of Afghanistan and Iraq. This situation questioned the neutrality of the United Nations humanitarian efforts when a member state was in violation of the UN Charter. It also saw the erosion of humanitarian space as the nature of warfare has shifted to involve non-state actors which fundamentally reject international norms and law, and humanitarians seen party to it. The early 2000s also saw the devastation brought on by the 2004 Indian Ocean Earthquake and Tsunami, which caused the international community to review its operational environment. In 2005 the United Nations launched the Humanitarian Reform Agenda which saw the establishment of the UN Cluster System which identifies lead agencies to lead sectoral requirements designated by the IASC.

The UN Cluster System was first implemented after the 2005 Pakistan Earthquake. A disaster most often requires militaries to work alongside civilian rescue and search teams, the general public, political decision-makers and humanitarian organizations represents opportunities for a unique civil–military interface. This can foster cooperation and a mutual transfer of values. Alternatively, it may result in competition, conflict and a clash of organizational cultures.[40] For the military, disaster response is included within the operational realm of HADR, which itself comes under Operations Other Than War (OOTW). In the defense policy realm HADR is often grouped as a 'shared interest' under Non-Traditional Security alongside counterterrorism, counter-piracy and anti-smuggling, cyberwar, military medicine, and peacekeeping. HADR is primarily centered around military engagement in responses to disasters caused by natural hazards in the Asia-Pacific.

However, HADR also includes military engagement in humanitarian assistance in conflict settings. In the Asia-Pacific, HADR remains primarily viewing through the lens of natural disasters as noted earlier. These contexts activate different actors or components of actors from different government line ministries to INGOs or in the case of the Red Cross and Red Crescent Movement, the International Committee of the Red Cross (ICRC) has the mandate for conflict settings whereas the IFRC has the mandate for disasters. However, both engage in either context where their niche capabilities are needed and their pre-existing relationships and presence in an affected country offer access to affected populations or national leadership of the overall disaster response effort.

The UN OCHA provides a platform for civil-military coordination (CMCOORD), a forum for dialogue and interaction between military and civilian actors in international humanitarian settings. The UN also provides guidance through a series of general operational handbooks for military engagement in disaster assistance such as the *UN-CMCOORD Field Handbook (2018)*, and *Humanitarian Civil Military Coordination— A Guide for the Military (2017)*. However, when military assistance in disaster relief is offered, militaries work directly with the host country to provide this. The United Nations mechanisms provide a link between militaries and the wider international humanitarian response to foster partnerships, minimize overlaps and fill gaps in the provision of humanitarian assistance. The Changi Regional HADR Coordination Centre is an effort to support an affected country's military with external militaries by providing a platform for the host military to identify their gaps and external militaries to support assistance in overcoming these. Importantly the UN framework is a dominant but not singular framework for the conduct of disaster response. More broadly in International Relations literature this is often reflected in calls for a more Global IR.[41] When it comes to the military, bilateral arrangements between an affected state and a foreign military play a decisive role in the conduct of HADR operations.

Disaster scenarios have many different actors involved to assist affected communities. The contention that militaries now engage in non-traditional security issues like disaster response will illicit some rejection from humanitarian actors as these military activities have been experienced worldwide since their founding. However, what has changed is the number of recognized (primarily civilian) actors in the field responding to non-traditional security issues and the levels of civil-military cooperation to address these concerns. There are divergent dominant opinions on the role of the military from Europe and the Asia-Pacific. In Europe, the notion that militaries are used "as a last resort" is highlighted in the Oslo Guidelines, whereas militaries being used as a "first responder" is recognized in the Asia-Pacific Conference on Military Assistance to Disaster Relief Operations (APC-MADRO) Guidelines. These guidelines highlight a greater ease with which countries in the Asia-Pacific engage their militaries and the last resort that European countries have in using their militaries. Gabrielle Simm argues that there are two conflicting documents over the role of the military in disaster response, notably the Oslo Guidelines (2007) and the APC-MADRO Guidelines (2010). However, the former guidelines reflect the practices in Europe while the latter reflect the practices in the Asia-Pacific.[42] In sum, the global humanitarian operational environment has evolved primarily focused on conflict settings whereas the experience in the Asia-Pacific has primarily focused on disaster settings. The variations in regional experience highlight the need for greater situational awareness among decision-makers.

Enablers—Multilateral and Bilateral Environment

In the previous two sections, the role of the military in disaster response is the result of myriad policy considerations including the scale of the disaster, the pre-existing relationship with the affected country, regional and global commitments, the direct and indirect costs involved in the deployment of the military, domestic political context and broader foreign policy goals. The decision, function and form of the military involvement in disaster response occurs within the international humanitarian system with relative regard for established norms and processes recognizing local and regional experience. This section explores the function and form of military involvement in disaster response and draws on humanitarian civil-military disaster partnerships as a concept to explain the mode of these partnerships.[43]

With a significant presence in HADR, militaries have sought to assess the standard operating procedures through which militaries are engaged in such efforts. In Southeast Asia, there are several ASEAN documents that provide for this guidance as the regional organization seeks to become a global leader in disaster management and emergency response by 2025.[44] These documents provide a clear indication of the role for militaries in disaster contexts, which focuses on asset deployment to bolster logistical capabilities to transport needed relief items and personnel into a disaster-affected country during the response phase and their engagement in pre-paredness activities such as contingency planning and scenario building and exercises like ARDEX. This sees foreign militaries working in part-nership with the national militaries and civilian components to achieve their objectives, which are often focused on limited-time, surge capacity through such endeavors as forming an airbridge to facilitate assistance provision in the aftermath of the 2018 Sulawesi Earthquake, Tsunami and Soil Liquefaction. The airbridge between Balikpapan and Palu was con-sidered the most efficient and safest way to reach Palu. It was operated with the support of 17 international assets under TNI leadership. Civilian organizations that actively engaged in dialogue with the military encoun-tered few constraints in transit and transportation of human and material resources.[45]

As disaster-affected countries build stronger institutional capacity to manage disasters and engage in greater civil-military partnership, the role of the United Nations itself has change to resemble more that of a clearing house than the most significant disaster responder. Even its capacity as a clearing house, this is now challenged in Southeast Asia by the ASEAN disaster management system, with the notable mandate afforded to the ASEAN Coordinating Centre on Humanitarian Assistance in disaster management (AHA Centre) as seen in the 2018 Sulawesi Earthquake, Tsunami and Soil Liquefaction. This response effort saw the AHA Centre provide the role usually assumed by UN OCHA and saw the UN coordi-nating mechanism sit within the AHA Centre Operations Centre.

The sub-region of the Asia-Pacific with the most active regional arrangements is Southeast Asia through the regional association ASEAN. Presently the ASEAN Regional Disaster Emergency Response Simulation Exercise (ARDEX) is the premier civilian and military exercise focused on HADR. There are also regular civilian and military meetings that cover HADR. For the civilian agencies, this is led by the ASEAN Committee on Disaster Management (ACDM). For the militaries it is through the ASEAN Defence Ministers Meeting (ADMM). The disaster response operational environment within ASEAN is civilian-led and military engagement supports the civilian structures of ASEAN Member States. However, in Myanmar the military forms a separate but equal structure under their 2008 constitution which makes for a more complex operating environment in disasters. Immediately prior to the referendum that confirmed the 2008 Constitution, Myanmar was struck by Cyclone Nargis then under its military junta, which saw a series of negotiations ensue to gain access to the disaster-affected population in the Irrawaddy Delta. This experience illustrated that whilst dominant international frameworks may be in place, the local context ultimately rules.

The military interest in investing time during 'peace time' to engage in exercises and forums with civilian agencies and NGOs in a conducive environment is to ensure that 'name cards' are exchanged before a crisis. It also allows the military to understand the decision-making structures and familiarize themselves with personnel of other organizations in the field.

The military-to-military coordination structure sees an array of compositions dependent on the presence/absence of pre-existing relationships. At the highest order of partnership these take the form of defense agreements in the form of Visiting Forces Agreements, formal membership of defense treaties or forums such as NATO, Five Power Defence Agreement, or the ASEAN Defence Ministers Meeting. In this highest order partnership, militaries often have formalized cooperation mechanisms that include HADR activities. They will have also moved from national leader-to-leader trusting exchanges to more systematic exchanges between militaries themselves. These partnerships often are characterized by high levels of trust between the militaries. At the most professionalized level at present these militaries often have well established positions of responsibilities for military engagement in HADR domestically. This often finds a civilian-led response with military support as is found in the militaries of the United States, Australia, New Zealand, UK, France and other European countries. This continues to dominate the discussions around disaster response at the global level.

Outside the global corridors of power, there are emerging alternative arrangements that reflect the domestic structure of the countries involved. These can often be equally trustworthy but less formal in their exchanges and more reliant on the relationships between individual national leaders and military chiefs. Through similar command-and-control structures of militaries there is a pre-existing formulation that allows for militaries to pinpoint their points of contact in partner militaries, these more often include liaison officers from a host military being assigned as the focal point in the disaster response partnership. In less formalized partnerships these liaison officers can be defense attaches in Embassies that ultimately need to go through high command rather than directly engaging the relevant division within the military. With such partnerships these relations are necessarily slower to achieve the desired effects as they go via the capitals than directly to the commanding officer in the affected area.

Oftentimes there are justifications for these partnerships such that the militaries involved are not yet as professionalized as other militaries or that there is more centralized control of the military by the civilian authorities or that the militaries hold a different function within the engaged country. The military's historic role within the country may involve more active participation outside what is often seen as the core functions of a professional military—warfighting and protecting national borders from external aggression. They may be involved in civilian affairs through their presence at the local community level such a township posts. These experiences therefore underline the need to understand the operational framework that exists in a disaster context to identify the function and form of the military response.

Conclusion/Emerging Issues

The humanitarian assistance and disaster relief operational environment has evolved significantly since the 2004 Indian Ocean Earthquake and Tsunami. The Asia-Pacific has witnessed the most substantive developments in disaster response with the increased formalization of the multi-stakeholder environment. Most notably for the region the 2008 ASEAN Agreement on Disaster Management and Emergency Response and the APC-MADRO Guidelines provide the dominant operational framework for military and civilian actors in Southeast Asia and the wider

Asia-Pacific. Some of the current challenges that face the HADR operational environment are the increasing likelihood that disaster-affected countries are increasingly invested in disaster preparedness and response activities particularly in the development of national processes to receive offers of assistance.

In Bangladesh, the Multi-National Coordination Centre (MNCC) standard operating procedures (SOPs) were drafted over two years from 2018 to 2019 with the support of Changi RHCC in Singapore. Disaster Response Exercise and Exchange Bangladesh (DREE) and Exercise Coordinated Response (Ex COORES) were jointly planned with the US and Singapore to test the SOPs. The Bangladesh Ministry of Disaster Management and Relief along with the Bangladesh Armed Forces Division, Singapore Armed Forces (SAF) and CFE-DM co-hosted Ex COORES in April 2019 in Singapore and the DREE took place in October 2019 in Bangladesh. The Armed Forces of the Philippines (AFP) have recently created three battalions of the army, navy and air force specially tasked with disaster response illustrating the direction of developments in disaster affected countries. Bangladesh and the Philippines highlight how countries exposed to natural hazards are developing their national capacities to respond to these challenges and offer broader insight into the future function and form of disaster governance in the Asia-Pacific.

Further within the global HADR operational environment there is momentum to recognize and support local humanitarian action. However, how national and foreign militaries will engage with local and devolved leadership remains an open question for individual national governments. It is often answered that their presence will be coordinated by the national government yet reliance on such answers highlights a gap in contingency planning for future disasters where national capacity is stretched to its limit and local leaders have the mandate to lead the humanitarian response. In the USA efforts to develop stronger civilian-military disaster preparedness at the local level have been undertaken and identified key challenges such as staff turnover, gaps in risk assessments and non-transferable research between sectors.[46] It is therefore incumbent upon decision-makers to consider ways to address these broad challenges as they will also be present in other countries. Of note is the need to shift the HADR operational environment away from a supply-driven model dominant at present where the focus is on providing HADR overseas based on internal considerations rather than on the needs of recipient countries.

The difficulty remains that those countries most exposed to natural hazards are often those with limited finances and are often already dependent on pre-existing relationships of another magnitude. It is therefore likely that the development of governance will continue to be a patchwork of structures with varying levels of capacity. The clearest example in the Asia-Pacific is the relationship between the Southwest Pacific Islands and Australia and New Zealand. For trusting relationships to develop between militaries and disaster-affected countries more generally there is a need for their participation which is often paid for by their development partners. Without such support then the individual country would be unable to participate in such exercises or forums to build the trusting relationships need for a more effective disaster response when in need. This dependence links back to larger political questions over the dominance of the United States, Australia, New Zealand, France and the UK within the Asia-Pacific and the variation in emerging capacity of countries exposed to natural hazards.

In some disaster-prone countries notably the Pacific Islands there are relatively few with a military—limited to Papua New Guinea, Fiji and Tonga. Outside these Pacific Island States, their security forces are police or paramilitary police forces. The increasing exposure to disaster and climate change (notably sea-level rise) have caused a shift in thinking within the dominant humanitarian civil-military architecture towards an operating model of civilian-military-police advocated by Australia and gaining traction within the wider Asia-Pacific on HADR. This is causing an evolution of the relationships between stakeholders to build trust with the relevant authorities in countries exposed to disasters on a more routine level during military exercises, simulations, and networking events. This reflects a growing shift in the recognition of partnerships being built between the countries concerned rather than a focus on supply-driven military response to a more nuanced matching supply with the needs of a disaster-affected country.

In terms of analytical framework, the HADR environment will engage several bids and influences for decision-makers to consider in HADR. The functional-humanitarian argument for military engagement in HADR remains the most prominent framework in the field of humanitarian affairs. This captures the differentiated interests of each stakeholder coalesce around HADR where the motivation of military engagement may well be to improve national security strategy and international security partnerships and alliances, but it can concurrently serve the needs of disaster-affected people. Rather it sees this coalescence as a coalition of stakeholders where each have their own constituencies but can cooperate to respond to a disaster.

Within the current global operational environment and indeed the emergent international cooperative HADR landscape focuses on developing civil-military partnerships. Within this framework civilian agencies, NGOs and militaries have established common ground through which to work together albeit at the lowest common denominator level. These manifest in the Asia-Pacific in AADMER and the APC-MADRO Guidelines. This international HADR landscape remains dominated by international actors including but not limited to International NGOs, the United Nations, and foreign militaries coordinating with one another to offer support to the host government increasingly in areas identified by them as requiring humanitarian assistance and disaster relief.

However, the domestic strategic considerations of foreign militaries engaging in disaster response are pertinent to understand the motivations behind such engagements. While the global commitments to humanitarian work remain the trigger to assist those in need, the function and form of military engagement draws on several domestic considerations. These domestic considerations include the opportunity for service personnel to engage in an active operational environment, commitments made in broader strategic partnerships and alliances, showcasing military capacity and force projection for both foreign and domestic audiences, and offering multi-functional options for governments facing financial constraints and scrutiny. As disaster affected countries across the Asia-Pacific develop their national capacities, these will become increasingly more important and will be the catalyst for a more significant shift from a supply-driven to a needs-based international humanitarian system.

Endnotes

[1] Dorothea Hilhorst and Greg Bankoff, "Mapping Vulnerability," in Gregg Bankoff, Georg Frerks and Dorothea Hilhorst (eds.), *Mapping Vulnerability: Disasters, Development and People* (London: Routledge, 2004). https://doi.org/10.4324/9781849771924.

[2] *ASEAN Agreement on Disaster Management and Emergency Response* (Jakarta: ASEAN Secretariat, 2008), p. 3.

[3] Elizabeth Ferris, "Future Directions in Civil-Military Responses to Natural Disasters," *Australian Civil-Military Centre Paper* 5, 2012, pp. 1–10.

[4] Elizabeth Ferris, "Future Directions in Civil-Military Responses to Natural Disasters," *Australian Civil-Military Centre Paper* 5, 2012, pp. 1–10.

[5] *Ibid.*, p. 993.

[6] Muthiah Alagappa (ed.), *Coercion and Governance: The Declining Political Role of the Military in Asia* (Stanford: Stanford University Press, 2001).

[7] Felix Heiduk, "From guardians to democrats? Attempts to explain change and continuity in the civil–military relations of post-authoritarian Indonesia, Thailand and the Philippines," *The Pacific Review*, vol. 24, no. 2, 2011, p. 253.

[8] Malesic, "The impact of military engagement in disaster management on civil–military relations," pp. 985–987.

[9] Annie Isabel Fukushima *et al.*, "Disaster Militarism: Rethinking U.S. Relief in the Asia-Pacific," *Foreign Policy in Focus*, 11 March 2014.

[10] Cynthia Enloe, "COVID-19: Turning Swords into Ventilators? Or is it Ventilators into Swords?" *Women's International League for Peace and Freedom*, 7 April 2020. https://www.wilpf.org/covid-19-turning-swords-into-ventilators-or-is-it-ventilators-into-swords/.

[11] Malesic, "The impact of military engagement in disaster management on civil–military relations," p. 987.

[12] Yusa Djuyandi, Heri Casnoto, and Wahyu Hidayat, "Military Operations Other Than War (MOOTW): Synergy of Indonesian National Armed Forces (TNI) and National Disaster Management Agency (BNPB) in Disaster Management," *Humanities & Social Sciences Reviews*, no. 4, 2019, pp. 111–121.

[13] Malesic, "The impact of military engagement in disaster management on civil–military relations," p. 993.

[14] David Capie, "The United States and Humanitarian Assistance and Disaster Relief (HADR) in East Asia: Connecting Coercive and Non-Coercive Uses of Military Power," *Journal of Strategic Studies*, vol. 38, no. 3, 2015, p. 321.

[15] Malesic, "The impact of military engagement in disaster management on civil–military relations," p. 986.

[16] Ferris, "Future Directions in Civil-Military Responses to Natural Disasters," p. 6.

[17] Vanessa Newby, "ANZUS cooperation in humanitarian assistance and disaster response in the Asia-Pacific: ships in the night?" *Australian Journal of International Affairs*, no. 74, no. 1, 2020, pp. 72–88.

[18] Fukushima *et al.*, "Disaster Militarism."

[19] Catherine Bragg, "International Humanitarian Assistance: What Must Change," *RSIS Commentaries No. 19042*, 11 March 2019. https://www.rsis.edu.sg/rsis-publication/nts/international-humanitarian-assistance-what-must-change/#.XyOCkigzbIU.

[20] Jesse M. Cunha, Yu-Chu Shen & Zachary R. Burke, "Contrasting the Impacts of Combat and Humanitarian Assistance/Disaster Relief Missions on the Mental Health of Military Service Members," *Defence and Peace Economics*, vol. 9, no. 1, 2018, pp. 62–77.

[21] Deon V. Canyon, Benjamin J. Reilly, and Frederick M. Burkle, "Military Provision of Humanitarian Assistance and Disaster Relief in Non-Conflict Crises," *Journal of Homeland Security and Emergency Management*, vol. 14, no. 3, 2017, pp. 1–5.

[22] *Ibid.*, p. 4.

[23] Julia Hornyacsek, "Why the Military Forces? The Role and Capabilities of Military Forces in the Protection Against Disasters," *Land Forces Academy Review*, vol. 23, no. 1, 2018, p. 12.

[24] Aida Idris and Saiful Nizam Che Soh, "Determinants of HADR Mission Success: exploring the experience of the Malaysian Army," *Disaster Prevention and Management*, vol. 23, no. 4, 2014, p. 456. https://doi.org/10.1108/DPM-01-2013-0003.

[25] Aruna Apte, Marigee Bacalod, and Ryan Carmichael, "Tradeoffs among Attributes of Resources in Humanitarian Operations: Evidence from United States Navy," *Production and Operations Management*, vol. 29, no. 4, 2020, pp. 1071–1090.

[26] Jeffrey Engstrom, "Re-conceptualizing the Role of the Military for International Disaster Relief in East Asia," in Alan Chong (ed.), *International Security in the Asia-Pacific* (UK: Palgrave MacMillian, 2018), pp. 401–417.

[27] Aruna Apte and Keenan D. Yoho, *United States Navy Humanitarian Assistance and Disaster Relief (HADR) Costs: A Preliminary Study* (Monterey, CA: Naval Postgraduate School, 2015), pp. 1–15.

[28] M. Factor, "What's the real defense budget?" Forbes, 28 March 2011. https://www.forbes.com/forbes/2011/0328/billionaires-11-capital-flows-mallory-factor-real-defense-budget.html?sh=1bb911e70da7.

[29] Apte and Yoho, *United States Navy Humanitarian Assistance and Disaster Relief (HADR) Costs.*

[30] Luis Ballesteros, Michael Yuseem and Tyler Wry, "Masters of Disasters: An Empirical Analysis of How Societies Benefit from Corporate Disaster Aid," *Academy of Management Journal*, vol. 60, no. 5, 2017, p. 1703. https://doi.org/10.5465/amj.2015.0765.

[31] Pablo Yuste *et al.*, "Synchronized Humanitarian, Military and Commercial Logistics: An Evolving Synergistic Partnership," *Safety*, vol. 5, no. 67, 2019, pp. 1–15.

[32] C.A. Hofmann and L. Hudson, "Military responses to natural disasters: Last resort or inevitable trend?" *Humanitarian Exchange Magazine*, no. 44, 2009, p. 1. Also see: Malesic, "The impact of military engagement in disaster management on civil–military relations," p. 983.

[33] Aaida Mamuji, "Canadian military involvement in humanitarian assistance: progress and prudence in natural disaster response," *Canadian Foreign Policy Journal*, vol. 18, no. 2, 2012, p. 217.

[34] *Ibid.*, p. 218.

[35] *Ibid.*, p. 219.

[36] A.J. Cunningham, *International Humanitarian NGOs and State Relations: Politics, Principles and Identity* (New York: Routledge, 2018), p. 23.

[37] Deon V. Canyon, Benjamin J. Ryan, Frederick M. Burkle, "Rationale for military involvement in humanitarian assistance and disaster relief," *Prehospital and Disaster Medicine*, 2019, pp. 2–3.

[38] David Fisher, "The Law of International Disaster Response: Overview and Ramifications for Military Actors," in Michael D. Carston (ed.), *Global Legal Challenges: Command of the Commons, Strategic Communications and Natural Disasters*, International Law Studies volume 83, 2007, p. 307.

[39] *Ibid.*, p. 308.

[40] Malesic, "The impact of military engagement in disaster management on civil–military relations," p. 981.

[41] Amitav Acharya, "Global International Relations (IR) and Regional Worlds: A New Agenda for International Studies," *International Studies Quarterly*, vol. 58, no. 4, 2014, pp. 647–659.

[42] Gabrielle Simm, "Disaster Militarism? Military Humanitarian Assistance and Disaster Relief." *Max Planck Yearbook of United Nations Law Online*, vol. 22, no. 1, 2019, pp. 347–375.

[43] A.D.B. Cook and S. Yogendran, "Conceptualising humanitarian civil-military partnerships in the Asia-Pacific: (Re-)ordering cooperation," *Australian Journal of International Affairs*, vol. 74, no. 1, 2020, pp. 35–53.

[44] *ASEAN Vision 2025 on Disaster Management* (Jakarta: ASEAN Secretariat, 2015).

[45] Angelo P. L. Trias and Alistair D. B. Cook, "Recalibrating Disaster Governance in ASEAN After the 2018 Central Sulawesi Earthquake and Tsunami," *RSIS Policy Report* (Singapore: RSIS, 2019), p. 10. https://www.rsis.edu.sg/wp-content/uploads/2019/12/PR191206_Recalibrating-Disaster-Governance-in-ASEAN_v3.pdf.

[46] Melinda Moore *et al.*, 2010. *Bridging the Gap: Developing a Tool to Support Local Civilian and Military Disaster Preparedness* (CA: Rand Corp, 2010).

https://doi.org/10.1142/9789811244292_0008

Chapter 7

Peace Operations and Military Organizations Between Internationalism and Statism

Hikaru Yamashita

Professor, Department of International Relations,
University of Shizuoka, Shizuoka, Japan

hikaru@u-shizuoka-zen.ac.jp

The relationship between military organizations and peace operations has been under constant debate since the birth of peacekeeping as we know it.[1] For instance, the frequently cited dictum commonly attributed (on somewhat hazy grounds) to former United Nations (UN) Secretary-General Dag Hammarskjöld (1953–1961)—"Peacekeeping is not a job for soldiers, but only soldiers can do it"—not only reveals a basic paradox surrounding military peacekeepers but also the fact that such perception has a rather long pedigree. Hammarskjöld is widely credited for declaring the basic principles that make up UN peacekeeping.[2] Since the days of Hammarskjöld at least, then, the use of the military for a type of international activity that came to be called "peacekeeping" has been accompanied by a sense of uneasiness for UN officials like Hammarskjöld as well as for, perhaps more commonly, defense officials and military planners.

But why does the military's participation into peacekeeping missions generate such uneasiness? Considering this question is made more complex by two broad, mutually related factors that emerged in recent periods. One is the fact that peacekeeping has undergone a dynamic process of expansion and evolution well beyond what Hammarskjöld himself might have expected. Whereas the majority of Cold War-era peacekeeping missions were intended to control the consequences of interstate wars especially during the critical transition phase from ceasefires to peace agreements, most of today's peacekeeping missions, deployed not just by the UN but also a growing number of regional organizations and ad hoc coalitions, are now multidimensional missions that are inserted into ongoing or recurrent intrastate wars and mandated to assist the country in a broad array of state/nation-building tasks within the framework of peace agreements. Using the typology introduced by Boutros Boutros-Ghali in the early 1990s, the phenomenon of multidimensional peacekeeping has come to include not just peacekeeping itself, but also all the other modes of UN peace operations: preventive diplomacy/conflict prevention, peacemaking and peacebuilding. Given this, and also because it is mostly in the context of peacekeeping that the role of the military poses the problem we just saw, this paper focuses on peacekeeping.

A second structural factor that complicates Hammarskjöld's dilemma is related to broader changes in international geopolitical relations in the post-Cold War world. From the notion of a "new international order" in the early 1990s through the "unipolar moment" in the early/mid 2000s to the "rise of the rest" in the 2010s,[3] the perception of international order has been changing very quickly. Indeed, the fact that the expression "post-Cold War" is still widely used to describe the current era suggests the fluidity of the situation: we know that we no longer live in the Cold War structure, but our sense of what kind of world we find ourselves in remains uncertain. The most recent grid of the debate on global order, also relevant for the theme of this chapter, is the rise of new power centres, particularly the People's Republic of China and Russia, and the resultant resurgence of inter-state competition on the global stage. Peacekeeping, as something founded on cooperation between state contributors, is bound to be affected by such emerging strategic trends.

One can therefore ask: did the dynamic changes in the nature of peacekeeping in the post-Cold War era help ameliorate the problem of uneasiness, or rather exacerbate it? And what does it mean for military organizations to engage in peace operations within the fast-changing strategic landscape of the 2020s? This chapter aims to explore these issues first by clarifying the nature of the dilemma associated with the military's role in peacekeeping. It then sketches how the relationship between military organizations and peace operations is being influenced by post-Cold

War changes both in peacekeeping and in international relations generally. The argument advanced here is that the relationship has become a complex one where competing rationales generate conflicting expectations for military peacekeepers and yet that the balance sheet appears to suggest that peacekeeping may be an increasingly tough sell for military organizations in the emerging security environment of this century.

Hammarskjöld's Dilemma

What makes peace operations difficult for military organizations? A useful starting point here is provided by the 2004 book *Forces for Good* edited by Lorraine Elliott and Graeme Cheeseman. Drawing from contributions to a conference in Canberra in late 2002, the book's authors highlight the rise of a cosmopolitan ontology that sees humanity as "bound together as a community of fate"[4] and explores, on the basis of this ontology, the idea of deploying military forces to defend the objectives, principles and rights of the global human community and their individual members. According to Elliott and Cheeseman, there is evidence that this envisaged role of military forces for a cosmopolitan ethic has found voice in international policy (e.g., the UN's 1999 report on the Srebrenica report, the responsibility to protect); and, yet, the key challenge is the gap between the conception and ethic of a global community and the real-life existence of nationally organized state militaries. That is, insofar as the ethic of cosmopolitan good assumes the existence of a cosmopolitan community as the first community of identity for its members, existent military organizations, founded on the logic and structure of national statehood, will need to be "internationalised" in order to fully embody and act on that ethic.[5] From this angle, therefore, the greatest obstacle for the realization of cosmopolitan objectives lies in the national foundation of the military organization itself. In order to work as "forces for good," they would have to be "qualitatively as well as materially different from traditional militaries in their identity and value structures."[6]

Peacekeeping, of course, is not a straightforward cosmopolitan project; but taking a cosmopolitan stand for the moment has the heuristic value of articulating Hammarskjöld's dilemma because it sheds light on the basic assumptions behind the institution of peacekeeping. These assumptions are as follows. On the one hand, peacekeeping was, and still remains, firmly rooted in interstate agreements. UN peacekeeping missions are established by UN Security Council resolutions on the basis of Chapters VI and VII of the UN Charter. As I have reviewed elsewhere,[7] many non-UN missions also routinely refer to relevant Security Council resolutions in addition to the decisions of the regional organizations that deploy such missions; and just like the UN, these regional bodies are also intergovernmental frameworks. Peacekeeping, in that sense, is no different from other types of multilateral operation whose effectiveness is dependent on the level and nature of interstate agreement and consensus. Peacekeeping missions are created, funded, staffed, equipped, deployed, directed and eventually terminated on the basis of decisions reached by sovereign national governments that comprise these frameworks.

On the other hand, however, peacekeeping, as a type of international activity, contains elements that are not easily reducible to national interests and interstate relations. Whether intended for the regulation and management of interstate conflicts (traditional missions), or as a tool for international intervention into and assistance for countries in civil war situations (multidimensional missions), peacekeeping needs to represent the "voice" of the international community at large in order to gain credibility and trust among the host population. This stance is partly based on the practical need to signal respect for the norm of non-interference and maintain a neutral (now impartial) stance vis-à-vis warring parties, but it is also anchored in the possibility—or hope—that peacekeeping organizers like the UN *can* embody the will of the international community. It is reasonable, therefore, to demand that military personnel employed by the UN should observe "full loyalty to the aims of the Organisation and to abstention from acts in relation to their country of origin or to other countries which might deprive the operation of its international character and create a situation of dual loyalty."[8] This conception of peacekeeping is apparent, for instance, in a radio address on 10 December 1956 by Hammarskjöld to the members of the UN Emergency Force (UNEF), whose creation led to the principles of peacekeeping. He told UNEF personnel that "You are solders of peace in the first international force of its kind ... You are the front line of a moral force which extends around the world, and you have behind you the support of millions everywhere."[9] Peacekeeping, in other words, is not just international but also internationalist in nature, at least in part.

Moreover, this aspect of peacekeeping appears to be strengthened with its post-Cold War evolution mentioned earlier. Contemporary peacekeeping is typically a Chapter VII operation deployed in the context of ongoing or recurring civil wars, often with grave humanitarian consequences. Authorized by Chapter VII provisions, peacekeeping thus became an important means for the realization of collective security. The logic of collective security, just like the one of peacekeeping, is based both on the solidarist ontology of "We the Peoples" against which various threats are identified, and on the capability and willingness by member states to forge a collective response. Importantly, the scope of the Charter's "threat to international peace and security" elaborated through UN Security Council resolutions has come to include various transnational security concerns including transnational organized crimes, terrorism, weapons of mass destruction, proliferation of small weapons, sexual violence, infectious diseases, humanitarian crises, massive human rights violations and, potentially, climate change.[10] In particular, human rights protection and humanitarian assistance have become staple tasks for many peacekeeping missions which are often mandated to use "all necessary measures" for these purposes.[11]

Authorized under the concept of collective security and specifically mandated to perform tasks that are partly justified using cosmopolitan language, the purpose and role of peacekeeping contains internationalist or solidarist elements that appear to be growing in significance. But when it comes to the means—in particular, *military* means—of peacekeeping, it is still overwhelmingly dependent on the will of sovereign states. It is worth remembering that the process of force generation for peacekeeping missions is driven by a sort of voluntarism. In the absence of "special agreements" between the Security Council and UN member states for the provision of armed forces in the service of international peace and security as stipulated in Article 43 of the UN Charter, states do not have the binding obligation to contribute their troops to UN-authorized activities including those under Chapter VII; and they have the right to agree (or disagree) with whatever requests the UN Secretariat might make to them. States, on their part, have their own interests and priorities which are predominantly defined in national rather than transnational terms and to which political leaders are held accountable. These considerations play powerful roles in states' decisions to participate in peacekeeping missions.[12]

In summary: Hammarskjöld's dilemma regarding military peacekeeping is rooted in the tension between internationalist or cosmopolitan ideals that characterize peacekeeping objectives on the one hand, and statist and nationalist frameworks within which military contributions to peacekeeping are decided upon on the other. Then, in what specific ways does this tension manifest itself?

Operational and Organizational Challenges

Participation in peacekeeping entails several complications and challenges for military organizations. They are related to the national strategy and policy of participating countries, to the way in which participation impacts the operation and organization of national militaries, and/or to the quality of deployed peacekeeping personnel. This section highlights some of these challenges as domains in which the uneasy relationship between peacekeeping and military organizations is concretized.

On the operational level, there is a basic tension between military efficiency and effectiveness on the one hand, and multilateral operations on the other.[13] Organizationally, the demand for the efficient and effective implementation of military operations is translated into the emphasis on interoperability and force cohesion through which effective command and control is ensured; and a degree of homogeneity among the deployed personnel is assumed in this setting. However, multinational operations, consisting of troop contributions from diverse participating countries, are by definition heterogeneous in personnel training and skills, logistics, equipment, language, operational culture, and risk perception. Achieving military efficiency in multinational operational environments is therefore difficult, if not impossible. This problem is somewhat mitigated in multinational coalition forces, especially when a willing and capable state takes the lead in the operation and provides the other participating countries with necessary resources: Australia in INTERFET (International Forces in East Timor) is a good example.[14]

But peacekeeping, organized through multilateral fora like the UN, is a different story. For one, peacekeeping missions are typically enabled by contributions from the member states of the organizing institution. Since these missions are instruments created by the decision-making body of that institution, special importance is attached to the multilateral nature of the missions. Gaining the widest possible participation from the members is important not just from the burden-sharing perspective but also for the sake of delivering the political message of global (or regional) solidarity to the warring parties. But the more multilateral an operation becomes, the more difficult it is for the operation to achieve effectiveness on the ground. A dilemma therefore exists between the operational requirements for effective interoperability and the political and normative demand for multilateral outlook.

This dilemma manifests itself in three specific areas. One area is in the generation of the peacekeeping force. Partly as a result of the aforementioned emphasis on multinational representation, there has been a steady increase in the number of countries contributing uniformed personnel to UN peacekeeping operations: 120 UN Member States contributed troops to the UN in June 2020, a nearly 35% increase from June 2000 (89).[15] On the other hand, what has remained constant is that the majority of the UN's uniformed peacekeepers—about 60% of troops and police personnel—still come from ten troop contributing countries (TCCs); the other contributing countries are providing "token" contributions of platoon-size units or less (individual officers). What tends to emerge as a result is that peacekeeping missions increasingly comprise a combination of relatively small unit-level contributions; and even these small units are sometimes made up of multinational contributions.[16] From a global, solidarist perspective, this seems all good. In the eyes of national military planners, however, this patchwork composition of peacekeeping forces casts doubts on the efficacy of peacekeeping as a military operation as it raises the question of coordination mentioned earlier. Moreover, this scepticism may be amplified by the potential impact of deployed capabilities on the national military. In the conventional logic of military organizations, its constituent formations (typically from the battalion level and up) are organized to perform assigned tasks in a more or less autonomous manner. They are therefore made up of not just combat units but also correspondingly sized support units that provide various services to the former.[17] But if one part of a military formation—a good example is logistics units that tend to be in greater need in contemporary peacekeeping[18]— deploys for international service for an extended period, it will likely affect the operational capability of the rest of the national force. These considerations likely lead to cautious attitudes on the part of national defense authorities towards new deployment requests.

A second concrete area in which we can see the same dilemma is the unique command and control structure of peacekeeping operations. Whereas multinational forces consist of national forces retaining national command and control, command and control in peacekeeping is a unique hybrid of national and international authorities. Peacekeeping missions are directed by the political body of the deploying organization at the ambassadorial or ministerial level that authorizes their initial deployment and regularly renews the mandates: the UN Security Council, the African Union's (AU) Peace and Security Council, or the Council of the European Union (EU) are prime examples. However, the day-to-day management of the missions is led by the administrative head of the organization (the UN Secretary-General, the EU High Representative, or the Chairperson of the AU Commission) who in turn delegates its authority to a (typically civilian) special representative. Although decided upon and politically directed by national representatives, the mandates, once granted, are seen as representing the will of the international community at large and the participating personnel are asked to behave in loyalty to the global or regional organization rather than their home government.[19] Here, the solidarist element of peacekeeping again clashes with the national will to control and protect its deploying personnel. From the perspective of national governments, they have underlying doubts about sending their citizens to missions that are perceived to be out of their reach. This is particularly so when the mission operates in fragile security environments and there is fear of serious security risks to the deployed personnel. Politically, such a scenario, if it materializes, can easily unleash a political crisis that pushes incumbent government leaders to the corner. This fear undermines governments' willingness to send their troops, and even if they agree to the deployment, they will try to curtail the potential risk (both in security and political terms) by adding conditions (caveats) to their use. However, just as in the case of the hodgepodge nature of peacekeeping force generation, the existence of national caveats greatly undermines operational effectiveness which in turn negatively influences states' perception of peacekeeping.

A third area in which we can see the dilemma revolves around the priority attached to peacekeeping training. As Diehl and Balas point out:

> Most soldiers receive extensive military training in basic combat skills. These may be fine for some peace operation activities on the coercive end of the scale. Yet other missions depend for their effectiveness on a complex set of what has been referred to as "contact" (more diplomatic) skills. For example, missions whose primary purpose is monitoring call for observational and analytical skills. Those that attempt to restore countries to functioning civil societies require a much broader range of skills, including interpersonal and intergroup relations, communication, negotiation, and, in the case of military operations, a mix of combat and

political skills. Organisational skills are needed for missions intended to limit damage, such as humanitarian assistance, or to rebuild institutions, infrastructure, and local economies, such as development assistance. A key question, however, is whether soldiers are actually being trained in contact skills.[20]

The challenge for military peacekeepers is therefore both to learn a new and potentially broad range of skills that are not covered in conventional military education and to apply these skills and modes of behavior in constantly shifting situations.[21]

Against this background, it makes sense for countries to create peacekeeping training centres and initiate joint exercises focused on peacekeeping scenarios with foreign partners; and many of them nowadays conduct some of these programmes often in cooperation with the UN and regional organizations. They are certainly useful in improving interoperability in multinational settings and sharing common standards and rules among future peacekeepers; they may also promise to instil the spirit of international solidarity and multinational cooperation. However, it should also be noted that these initiatives do not necessarily ensure a stable supply of military peacekeepers. The problem here is twofold: (1) the gap between peacekeeping trainees and the actual peacekeepers, and (2) the reluctance to create within the ranks a military specialization for peacekeeping missions. Since many peacekeeping training sessions and exercises take place outside the actual mission context (e.g., at training centres and military schools), there has been a recognizable gap between the recipients of these training opportunities and those who get actually deployed.[22] Such gaps would disappear if countries decide to institutionalize a group of "peace operations specialists" that are specifically trained for peacekeeping and other related missions, able to deploy at short notice, and capable of leading the national contingents. From a globalist perspective, this appears to be a logical option.[23] But in reality, as far as this author is aware, the idea of peacekeeping specialization has not yet been in existence even in developed countries. Clearly, this is partly a matter of resources and partly a question of commitment. Creating and keeping a stream of military specialists for international service obviously requires investments in financial and human resources terms; and this in turn cannot be justified without a strong political consensus on the desirability of contribution to global causes such as peacekeeping. The limited progress towards a peace operations speciality within national militaries indicates the weakness of such consensus.

Two points are in order at this juncture. Firstly, the expansion of peacekeeping in recent decades tends to compound these challenges. The expanded mandate of peacekeeping demands a wider base of troop contributors, giving the peacekeeping force a more globalist outlook in theory but also a difficult task of intra-mission coordination in practice. The command and control problem will be similarly affected by the increased number of state contributors as well as the involvement of multiple civilian sectors that is frequently required in multidimensional missions. The growing complexity of the mandates will set the bar for peacekeeping training higher as more sophisticated "contact skills" are expected of military peacekeepers.

A second point worth making here is that the three organizational challenges all boil down to strategic debates where the solidarist-versus-nationalist schema has more direct implications. From the national perspective, underlying these challenges is the question as to what extent the government should balance its contribution to global peace operations vis-à-vis the other international and national activities. Insofar as the operational and the strategic are thus closely connected in Hammarskjöld's dilemma, it is important to explore what this dilemma signifies on the political level.

Political and Strategic Challenges

On the political and strategic level, the tension between nationalism and cosmopolitanism that constitutes Hammarskjöld's dilemma is translated into the question of how to strike a right balance between national defense and international cooperation. These two considerations are of course not necessarily incompatible with each other. Cooperation with other states and partners on a bilateral or multilateral basis can and indeed does contribute to improved national defense—utilizing the UN's collective security arrangement in dealing with external aggressors (e.g., the Gulf War) is a good example. But they may also work against each other when they point towards different directions while demanding the same resources. In this case, prioritization becomes necessary.

Generally speaking, the extent and nature of the relationship between national defense and international cooperation is dependent on two factors: (1) the country's perception of external threats, and (2) the strength of "cooperative" strategic culture. As for the latter factor, some countries have a long tradition of commitment to international peace which is seen as part of their identity and national interest. According to Bergman, for instance, in Nordic states "there is general agreement that the duties of their militaries transcend national borders."[24] They see their involvement into peacekeeping "as much an expression of their commitment to international peace as their support for the values underpinning the United Nations."[25] Following this self-image, participation into peacekeeping is projected as "an expression of solidarity with innocent victims of war and conflict."[26] Countries with such strategic culture tend to maintain an active posture towards peacekeeping, although the way in which they do so varies.[27] And yet even these countries cannot ignore the (perceived) reality of threats and challenges to their own security and defense, and when the country's sense of insecurity is heightened, international cooperation is less prioritized in favor of a more conventional, counter-threat posture.

Then, what is today's security landscape, and what implications does it have for national engagement in cooperative global endeavors like peacekeeping? In view of the rapidly changing nature of the current security environment the picture is admittedly very sketchy; nations also have diverse threat perceptions that defy easy generalization. But various accounts seem to suggest that the contemporary world of security is uniquely characterized by two rising and interconnected elements: (a) the rise of new power centres that challenge the position and values of established powers, and (2) the emergence of new security domains that provide new sites of influence and power projection.

As Tardy pointed out as early as 2012,[28] the rise of new powers does not necessarily lead to their posing a fundamental challenge to the existing regime of global peacekeeping. While asserting modifications in some of its normative elements such as human rights and the responsibility to protect,[29] these states generally accept or even promote the current regime of peace operations as long as it serves their growing global stature and influence. After all, China and Russia are the permanent members of the UN Security Council. China is now the second largest financial contributor to the UN's peacekeeping budget as well as a top ten troop contributor. India has long been one of the most ardent supporters of UN peacekeeping, while Brazil played a pivotal role in peacekeeping in the Western hemisphere (Haiti in particular).[30] The summit meetings of the BRICS grouping (Brazil, Russia, India, China and South Africa) have consistently projected the UN and peacekeeping in a positive light.[31] From their perspective, the UN and other intergovernmental organizations are seen as reform-able and thereby potentially useful resources for their version of a multipolar order.

Rather than making peacekeeping a direct arena for major power contest, the rise of new powers influences peacekeeping by changing other states' security calculus. One area in which such influence is being felt is regional and intrastate conflicts, the type of conflicts to which peacekeeping has been actively mobilized in recent periods. In the face of these conflicts there is a growing fear that they might become sites in which major power interests clash in the way they did in the Cold War years. The conflict in Syria from 2011 provides a good illustration of how changed geopolitical relations affect the chances for peacekeeping deployment. A brief comparison with two recent conflicts in Syria's vicinity may be worth mentioning here: Kosovo and Libya. The Kosovo conflict of 1999 saw the deployment of UN, NATO (North Atlantic Treaty Organisation), EU and OSCE (Organisation for Security and Co-operation in Europe) missions despite political and normative disagreements between the West and Russia and China (the latter's embassy in Belgrade was mistakenly bombed by NATO forces). In Libya in 2011, the UN Security Council managed to agree on the Chapter VII authorization for NATO's mission, again with Russia's and China's acquiescence in what was justified by its proponents as a "humanitarian" mission. In Syria, however, Russia, with China's support, was determined not to allow another Western intervention, repeatedly vetoed draft resolutions for sanctions against the Syrian government, and eventually started its own military intervention to prop up the al-Assad regime. As a result, only a short-lived observation mission was authorized by the Security Council. The Russian and Chinese position remained essentially the same throughout these years—they were against forcible intervention into other states (except perhaps for their own), and supported different local political groups than those favored by Western countries. What differed was the determination to push their demands through. With the US, Iran, France, UK, Turkey and Arab countries providing direct or indirect military support to different groups in the country, the war in Syria became what was often a common phenomenon in the Cold War period: proxy war. If such conflicts increase in the future, there will be correspondingly less chance for peacekeeping engagement.

In general terms, the phenomenon of emerging powers is associated with more assertive actions towards other states and the launching of aggressive foreign policy initiatives—China's actions in the South

China Sea and its Belt and Road Initiative, or Russia's military operations in Ukraine and Georgia are obvious examples. These actions, often involving the use of military force and the formation of new or reinforced security cooperation with junior partners, naturally put other major powers and neighboring states on alert, making them re-evaluate their policies in preparation for scenarios of heightened tension with the emerging powers. Prospects for great power competition also loom large in the global economy where established powers try to protect their (declining) industrial and technological advantages against rising powers. It is not difficult to see how the thus revived interest in "geo-economics" combines with increased geopolitical concerns to reinforce the perception of contemporary international politics as a competitive rather than cooperative game that tests the overall power capabilities of major powers.[32] Indeed, a perusal of recent commentaries suggests that the fusion of geopolitics and economic (and technological) competition appears to re-establish itself as the predominant (if not the only) angle from which to debate the relationship among the United States, Russia, China, EU and beyond.[33]

Moreover, this revival of great power competition appears to be reinforced with another trend in international security: the emergence of new security domains and issues. The creation of cyber space and the rapid proliferation of automation and telecommunications technologies over the past two decades are all advances that are powerfully driven by commercial interests. But as the internet, network-based communications and artificial intelligence (AI) are becoming the essential parts of economic and social infrastructure as well as the core components of military capabilities, the question of which country will achieve predominance in these areas has strong security implications.

In power political terms, the emergence of these domains means that they provide new sources of power projection, and therefore new sites of competition among established and rising powers. Emerging powers, notably China, are rapidly catching up on this front. For instance, Allison and his co-author warn that China is already "a *full-spectrum peer competitor* of the United States in commercial and national security applications of AI" and will overtake the US in the near future unless the latter mobilizes a serious national effort.[34] China is also said to out-innovate other nations in other areas of new technology including 5G networks, autonomous weapons, public surveillance, electronic finance, biotechnology and space.[35] By creating new paths to global influence and power, the emergence of new industries and technologies is likely to intensify a new game of great power contest.

For the purpose of this paper, it is worth emphasizing that these technological advances have implications for most parts of our society whose security is increasingly dependent on internet connectivity and network-based data transmission; and military organizations are no exception as network centrality has become integral to their operations and capabilities, potentially including those using nuclear weapons.[36] For military organizations, therefore, the current security landscape underlines the importance not only of military preparedness in the case of potential contingencies involving emerging powers but also of upgrading their military capabilities to incorporate latest technological innovations. Competition in the military domain becomes increasingly technological, demanding a greater portion of the defense budget as a result. And it is within these broader strategic trends that debates about future participation into peace-keeping missions will unfold in many countries.

Conclusion

In this chapter I argued that the uneasy relationship between the military and peace operations, encapsulated in Hammarskjöld's dictum, is understood in terms of the basic tension between the nation state-based logic of military organization and the solidarist ideal of peacekeeping. On the operational level Hammarskjöld's dilemma involves competing demands of military effectiveness and multinational cooperation that are observed in the areas of peacekeeping force generation, command and control, and training priorities. And these challenges are in turn connected to and compounded by recent strategic trends (rising powers and new security domains) which place states in conditions of global competition rather than cooperation.

Then, what chances are there for the military's role in peace operations in the future? The foregoing analysis suggests a rather bleak future. At least three broad scenarios suggest themselves. One is the possibility that peace operations might revert to the Cold War practice of sending traditional missions in response to a limited number of mostly interstate conflicts. If major powers are preoccupied with their own strategic competition and if, as a result, more and more conflicts (both intrastate and interstate) become proxy wars, the space for peace operations will shrink correspondingly. In this scenario peace operations will be limited to conflicts taking place in the border zones between great powers' spheres of influence, or those that are of marginal interest to them and yet are seen as requiring some sort of international action. Given that the number of conflicts falling into this category is small, however, this pattern will likely lead to an observable decrease in peace operations. Also, since the type of missions that is likely to fit this pattern will be a traditional one, the post-Cold War legacy of multidimensional peacekeeping may be gradually lost.

A second possibility is to use peace operations as a means of strategic competition between major powers. Although, as pointed out earlier, peacekeeping has so far avoided becoming a site of competition between major powers, the conditions of global competition may push them towards a more competitive use of peace operations whereby they manage conflicts in their zones of influence or even try to intervene into ones taking place in the rival's spheres. While this might drive the demand for peace operations into the future, the activity may change its character—while, as argued above, peace operations are a cooperative endeavor founded in part on global solidarism, in this scenario it may end up being a mode of intervention to promote the intervener's interests against its rivals. In short, peace operations will lose its globalist credentials.

A third possibility is to envisage peace operations as an "enlightened" means of strategic cooperation between major powers. As mentioned already, there appears to be broad consensus on the value of peace operations. Major powers may build on this global consensus to use peace operations as an area where they can foster mutual confidence by developing contacts among security officials and engaging in field activities together. In other words, this suggests the use of peace operations in two senses—among parties to the conflict, and among the mission's contributing countries. Clearly, however, this scenario has its own limitations. Generally, it goes against the emerging pattern of major power interaction we already saw, and it is questionable to what extent such cooperative engagement by means of peace operations can contribute to confidence building among major powers. Also, it requires a quality leadership in major powers as well as others that can balance out the cost-benefit calculation of strategic cooperation and that of strategic competition.

The reality will be some combination of all three scenarios, and yet the extent to which one scenario will predominate over the others will have important implications for the broad contour of peace operations.

This is so particularly because the first two of the scenarios involve a fundamental change in the nature of peace operations. And each scenario suggests a different terms of participation for national contributors—that is, they may be able to limit their contribution under the first scenario; they can or must be selective in choosing partners and areas of deployment under the second scenario; or they may be asked to contribute what they can within a broad range of mission mandates under the third scenario. On the other hand, however, it is worth reiterating that the practice of peace operations eventually depends on how actors (especially states) use it. As hinted in the last paragraph, what may matter in steering the future course of global peace operations is an imaginative leadership to tap into the potentials of this activity in such a way that brings security and stability among the powerful as well as conflict-prone members of international society.

Endnotes

[1] This chapter focuses on peacekeeping as developed by the UN and regional organizations since the Cold War years. It should be noted, however, that there were a few cases of multilateral operations in the pre-UN era that could meet the contemporary definition of peacekeeping. For an excellent account of peacekeeping practices before the Second World War, see Norrie MacQueen, *Peacekeeping and the International System* (Abingdon: Routledge, 2006), Ch. 2.

[2] *Summary Study of the Experience Derived from the Establishment and Operation of the Force: Report of the Secretary-General*, UN Doc. A/3943, 9 October 1958.

[3] See, for instance, Address Before the 45th Session of the United Nations General Assembly in New York, New York, 1 October 1990 (https://www.presidency.ucsb.edu/node/264816, accessed 7 January 2021); Charles Krauthammer, "The Unipolar Moment," *Foreign Affairs*, vol. 70, no. 1, 1990/1991, pp. 23–33; Charles Krauthammer, "The Unipolar Moment Revisited," *The National Interest*, no. 70, Winter 2002/2003, pp. 5–18; Fareed Zakaria, *The Post-American World: And The Rise Of The Rest*, updated and expanded edition (London: Penguin, 2011); Joseph S. Nye, Jr., *Is the American Century Over?* (Cambridge: Polity, 2015).

[4] Lorraine Elliott and Graeme Cheeseman, "Introduction," in Lorraine Elliott and Graeme Cheeseman (eds.), *Forces for Good: Cosmopolitan Militaries in the Twenty-first Century* (Manchester: Manchester University Press, 2004), p. 2.

[5] *Ibid.*, pp. 4–5.

[6] Lorraine Elliott, "Cosmopolitan Ethics and Militaries as 'Forces for Good,'" in Elliott and Cheeseman (eds.), *Forces for Good*, p. 24 et seq.

[7] Hikaru Yamashita, *Evolving Patterns of Peacekeeping: International Cooperation at Work* (Boulder: Lynne Rienner, 2017), Ch. 5.

[8] A/3943, para. 168.

[9] Quoted in Roger Lipsey, *Hammarskjöld: A Life* (Ann Arbor: University of Michigan Press, 2016), p. 311.

[10] Hikaru Yamashita, "Reading 'Threats to International Peace and Security,' 1946–2005," *Diplomacy & Statecraft*, vol. 18, no.3, September 2007, pp. 551–572; Security Council Resolutions 1308 (17 July 2000), 1820 (19 June 2008), 2117 (26 September 2013), 2177 (18 September 2014), 2195 (19 December 2014); SCOR, 8451th meeting, UN Doc. S/PV.8451, 25 January 2019.

[11] Although the Security Council prefers "measures" to "means" in recent resolutions, in practical terms they share the same effect of authorizing the use of force to UN operations. Niels Blocker, "Outsourcing the Use of Force: Towards More Security Council Control of Authorized Operations?" in Marc Weller (ed.), *The Oxford Handbook of the Use of Force in International Law* (Oxford: Oxford University Press, 2015), p. 213.

[12] For a comprehensive review of national motivations behind peacekeeping contributions, see Alex J. Bellamy and Paul D. Williams (eds.), *Providing Peacekeepers: The Politics, Challenges, and Future of United Nations Peacekeeping Contributions* (Oxford: Oxford University Press, 2013), in particular Ch. 19 by the editors.

[13] Alan Ryan, "Cosmopolitan Objectives and the Strategic Challenges of Multinational Military Operations," in Elliott and Cheeseman (eds.), *Forces for Good*, pp. 68–72. For a more general discussion of barriers to the military effectiveness of peacekeeping, see, e.g., Mats Berdal, "What Are the Limits to the Use of Force in UN Peacekeeping?" in Cedric de Coning and Mateja Peter (eds.), *United Nations Peace Operations in a Changing Global Order* (Cham: Palgrave Macmillan, 2019), pp. 118–123.

[14] For a good account, see, e.g., Peter Londey, *Other People's Wars: A History of Australian Peacekeeping* (Crows Nest: Allen & Unwin, 2004), Ch. 15.

[15] Data from the UN Department of Peace Operations website https://peacekeeping.un.org/en/troop-and-police-contributors, accessed 6 January 2021. I quoted the figure in June 2000 because the current surge in UN peacekeeping started from this period.

[16] See, e.g., Donald C.F. Daniel, Paul D. Williams, and Adam C. Smith, "Deploying Combined Teams: Lessons Learned from Operational Partnerships in UN Peacekeeping," International Peace Institute, August 2015.

[17] In the case of infantry, for instance, the proportion of non-combat units becomes larger with the size of the unit. James F. Dunnigan, *How to Make War: A Comprehensive Guide to Modern Warfare in the 21st Century*, fourth edition (New York: Quill, 2003), p. 18.

[18] Susan Smith, "Logistics and Multinational Military Operations," in Elliott and Cheeseman (eds.), *Forces for Good*, pp. 87–88.

[19] In UN peacekeeping, this is apparently one of the first lessons that trainees learn in pre-deployment training. See Integrated Training Service, UN Department of Peace Operations, *Core Pre-deployment Training Materials for United Nations Peacekeeping Operations* (New York: United Nations, 2017), Module 1, Lesson 1.1; see also Module 3, Lesson 3.1.

[20] Paul F. Diehl and Alexandru Balas, *Peace Operations*, second edition (Cambridge: Polity, 2014), p. 212; see also Dunnigan, *How to Make War*, p. 465.

[21] Diehl and Balas, *Peace Operations*, p. 213.

[22] See, e.g., Chr. Michelsen Institute in association with Itad, *Building Blocks for Peace: An Evaluation of the Training for Peace in Africa Programme*, Norwegian Agency for Development Cooperation (Norad) Report 6/2014, November 2014, Chs. 3-4.

[23] For proposals, see, e.g., H. Peter Langille, *Bridging the Commitment-capacity Gap: Existing Arrangements and Options for Enhancing UN Rapid Deployment*, The Center for United Nations' Reform Education, Monograph No. 19, November 2002; Robert C. Johansen (ed.), *A United Nations Emergency Peace Service to Prevent Genocide and Crimes against Humanity* (New York: World Federalist Movement Institute for Global Policy, 2006).

[24] Annika Bergman, "The Nordic Militaries: Forces for Good?" in Elliott and Cheeseman (eds.), *Forces for Good*, p. 171.

[25] *Ibid.*

[26] Norwegian Ministry of Foreign Affairs, *The Nordic Countries and International Peacekeeping*, p. 1, quoted in Bergman, "The Nordic Militaries," p. 175.

[27] For instance, Nordic states used to be leading contributors to UN peacekeeping during the Cold War years, but their focus shifted to NATO and/ or EU operations from the early 2000s. Peter Viggo Jakobsen, "The Nordic Peacekeeping Model: Rise, Fall, Resurgence?" *International Peacekeeping*, vol. 13, no. 3, September 2006, pp. 381–395. See also the special edition ("A European Return to UN Peacekeeping?") of the same journal (*International Peacekeeping*, vol. 23, no. 5, November 2016.

[28] Thierry Tardy, "Emerging Powers and Peacekeeping: an Unlikely Normative Clash," GCSP Policy Paper 2012/3, Geneva Centre for Security Policy, March 2012 (https://www.files.ethz.ch/isn/141118/Emerging%20Powers%20and%20 Peacekeeping.pdf), accessed 7 January 2021.

[29] Richard Caplan, "Peacekeeping in Turbulent Times," *International Peacekeeping*, vol. 26, no. 5, November 2019, p. 528.

[30] See, e.g., Moritz Schuberth, "Brazilian Peacekeeping? Counterinsurgency and Police Reform in Port-au-Prince and Rio de Janeiro," *International Peacekeeping*, vol. 26, no. 4, August 2019, pp. 487–510.

[31] Malte Brosig, "Ten Years of BRICS: Global Order, Security and Peacekeeping," *International Peacekeeping*, vol. 26, no. 5, November 2019, p. 523.

[32] Michael Lind, "The Return of Geo-economics," *The National Interest*, no. 164, November/December 2019, pp. 12–18.

[33] National Security Strategy of the United States of America, December 2017, pp. 2–3; Brian D. Blankenship and Benjamin Denison, "Is America Prepared for Great-power Competition?" *Survival*, vol. 61, no. 5, October-November 2019, pp. 43–64; Mark Leonard, *et al.* "Securing Europe's Economic Sovereignty," *Survival*, vol. 61, no. 5, October–November 2019, pp. 75–98.

[34] Graham T. Allison and "Y," "The Clash of AI Superpowers," *The National Interest*, no. 165, January/ February 2020, pp. 11–24 (quotes from p. 11).

[35] For a comprehensive coverage, see in particular Brooking's Global China project (https://www.brookings.edu/research/global-china-technology/, accessed 7 January 2021).

[36] See, e.g., David C. Gompert and Martin Libicki, "Cyber War and Nuclear Peace," *Survival*, vol. 61, no. 4 (August–September 2019), pp. 45–62.

Index

A

action-reaction, 59–60, 76, 91
Afghanistan, 48–49, 68, 75, 88, 94,
  96–97, 116, 119
Al Qaeda, 89, 92–93, 97
arms control, xix, xxv
arms dynamic, xx, xxix, 57, 59–60,
  63, 76
arms racing, xix–xx, xxix, 59, 71
artificial intelligence (AI), 66, 143
ASEAN Agreement on Disaster
  Management and Emergency
  Response, 107–108, 123–124
ASEAN Coordinating Centre on
  Humanitarian Assistance, 122
ASEAN Defence Ministers Meeting,
  145
Association of Southeast Asian
  Nations (ASEAN), xix, 5, 105, 122
Australia-New Zealand-United States
  (ANZUS), 122

B

battle, centrality of, xxi–xxii, xxv, 44,
  51, 68, 91, 97
Battle of Singapore, 9, 42–43, 50
Belt and Road Initiative, 143

C

character of war, xxix, 82–84, 89,
  94–95
Clausewitzian paradigm of strategy
  and war, xix, xxi, 51, 82–83,
  92–94, 96
Cold War, xvii, xxiii–xxv, xxvii, 3, 8,
  42, 62–63, 69, 73, 116, 132,
  142–143
Cold War, post-, 85, 117, 119, 134,
  145
conflict zone, 12, 89–90, 108,
  117–118, 145
conscription, 33, 46
conventional capability/force
  structure, xxiii, 114, 141
conventional deterrence, xxvii
conventional operations/strategy,
  xxiv–xxv, 8, 71–73, 94, 137
counterterrorism, xxviii, 32, 59,
  72–73, 120
cyber, xviii, 70, 87, 93, 120, 143
cyclone, 72, 105–107, 123

D

defense policy, xx, 9–12, 19, 21–22,
  24, 26, 32, 47

demilitarized zone, 5
deterrence, xix–xx, xxiii, xxv–xxvii,
    46, 91
digital communications, 60, 62, 67,
    84, 97
disarmament, xix, xxv, 63, 65, 74, 76

E
ethnocentrism, 22

F
Five Power Defence Agreement, 123
flash mob, 82, 90, 92–93
foreign fighters, 95–98, 102
France, 7, 61, 63, 91, 123, 126, 142

G
geography, military and political,
    xxiv, xxviii, 1–4, 6–13, 19, 21, 23,
    25, 34
geopolitics, classical, xvii, xxviii,
    2–7, 10–11, 143
geopolitics, critical, 2, 11
globalism, 139–140, 145
globalization, 47, 61, 71

H
Hezbollah, 89, 93
humanitarian assistance and disaster
    relief operations, 48, 59, 72–73,
    105–110, 112–113, 115–122,
    124–127
hybrid threats, 94, 113, 138

I
Indonesia, xix, 10, 41–42, 45, 48–49,
    61, 63–65, 72–73, 76, 111
information, strategic aspects of, 27,
    32, 67–68, 88, 90–91, 95, 98
information technology, 60, 62, 66,
    69, 71, 82–84
Iraq, 48, 61, 68, 75, 81, 87–88, 94,
    96–97, 119

Islamic State, 81–82, 86–89, 96–97
Israel, 8, 10, 27, 45–46, 61

J
Japan, 3, 7–9, 30–31, 33, 41–44, 72,
    98, 105–106, 114

K
Korea, North and/or South, 3, 5–6,
    13, 15, 30, 115
Korean War, 91, 97
Kosovo, 91, 142

L
lone wolves, 82–83, 87

M
Malaysia, xix, 3, 10, 43, 45, 63–65,
    72–76,
maritime environment, 3, 45–48, 63,
    74
metaphor, strategic utility of, 23, 25,
    82–83, 89–94, 96
military history, xxii–xxiv,
    xxviii–xxix, 4, 11, 19, 21, 25, 29,
    42–52
military modernization, xviii, 58–59,
    64, 67, 69, 72–75, 77, 91
military power, xviii, xx–xxi,
    xxiii–xxv, 2, 13, 17, 23, 46, 59, 68,
    72, 93
military technology, xxii–xxiii, xxx,
    48, 60–61, 64, 74
Myanmar, xix, 63, 65, 72, 105, 107,
    123

N
nation, imagined community, 85,
    89–90
New Zealand, xviii, 26, 48, 112, 123,
    126
non-traditional security, 93, 110, 114,
    120–121

North Atlantic Treaty Organization (NATO), 8

nuclear weapons/strategy, xviii, xxiii–xxiv, xxvi–xxvii, 8, 21–22, 30

O

offence-defense theory, xx

operations other than war (OOTW), xxvii–xxviii, 75, 113, 120

order of battle, 33–34

P

peace operations, xxviii–xxix, 58, 120, 132–142, 145–146

People's Republic of China, xviii, 3–4, 62–63, 75–76, 114, 132, 141–143

Philippines, xix, 63–65, 75, 106–107, 112–113, 125

R

Red Crescent/Red Cross, 105–106, 108–109, 117, 119–120

Revolution in Military Affairs (RMA), 62

Russia, 7, 58, 61–64, 70, 72, 75–76, 132, 141–143

S

security community, 31

security dilemma, xviii

Singapore, xvii, xix, xxix, 3, 7, 9–11, 25, 33, 42–52, 61, 63, 65, 73–76, 125

Sino-US competition, 3

Soviet Union, xxiii, xxv–xxvi, 21–22, 24, 28, 31, 62, 81, 88

Stockholm International Peace Research Institute (SIPRI), 61

strategic culture, 20–30, 31–33, 68, 141

strategic studies, definition of, xxi–xxv

strategic success, xxvii, 32, 68, 92, 96–98

strategy, definition of, xx–xxi, xxiv, xxx, 31, 92

street brawl, war as, 82, 92–94, 96–98

Syria, 81, 87–88, 96, 142

T

Taliban, 89, 93

terrorism, xviii, xxii, xxiv, 71

Thailand, xix, 63, 65, 72, 75–76

Total Defence, 33, 45

U

United Kingdom, 61, 123, 126, 142

United Nations Department of Peacekeeping Operations, xxviii

United Nations Emergency Force, 134

United Nations Office for the Coordination of Humanitarian Affairs, 119–120, 122

United Nations Security Council, 119, 134–135, 138, 141–142

United States, xxvi–xxviii, 3–4, 7, 13, 21–22, 30–32, 58, 61–64, 67–70, 72–73, 76, 91, 93–94, 98, 112, 115, 119, 123, 126, 143

V

Vietnam, xix, 30, 41–42, 45, 63, 65, 91, 97–98

Vietnam War, 30, 41–42, 45, 91, 97–98

vulnerability, strategic narrative, 3, 5, 33, 45, 47, 52, 107

W

Warsaw Pact, xxiv, 8, 62–63, 75, 78

World War II, Asia-Pacific, 3, 7–8, 33, 41–42

Z

zweikampf, 97–99